KV-435-205

ML

# Discrete electronic components

# Discrete electronic components

F. F. MAZDA
*Manager, Component Engineering*
*Rank Xerox Ltd*

CAMBRIDGE UNIVERSITY PRESS

*Cambridge*
*London   New York   New Rochelle*
*Melbourne   Sydney*

Published by the Press Syndicate of the University of Cambridge
The Pitt Building, Trumpington Street, Cambridge CB2 1RP
32 East 57th Street, New York, NY 10022, USA
296 Beaconsfield Parade, Middle Park, Melbourne 3206, Australia

© Cambridge University Press 1981

First published 1981

**Printed in the United States of America**

*British Library Cataloguing in Publication Data*
Mazda, F. F.
Discrete electronic components
1. Electronic apparatus and appliances
I. Title
621.3815    TK7870    80-42041
ISBN 0 521 23470 0

# Contents

| | | |
|---|---|---|
| Preface | | page vii |

## 1. Discrete semiconductors
| | | |
|---|---|---|
| 1.1 | Introduction | 1 |
| 1.2 | Semiconductor production techniques | 1 |
| 1.3 | Diodes | 3 |
| 1.4 | Transistors | 11 |
| 1.5 | The thyristor | 19 |
| 1.6 | The triac | 22 |
| 1.7 | The gate turn off switch | 23 |
| 1.8 | Trigger devices | 24 |

## 2. Optoelectronic components
| | | |
|---|---|---|
| 2.1 | Introduction | 28 |
| 2.2 | Optical terminology | 28 |
| 2.3 | Optical sources | 30 |
| 2.4 | Optical detectors | 34 |
| 2.5 | Optical couplers | 38 |
| 2.6 | Displays | 40 |
| 2.7 | Optical communication | 49 |
| 2.8 | Holography | 51 |

## 3. Resistors
| | | |
|---|---|---|
| 3.1 | Introduction | 52 |
| 3.2 | Resistor principles | 52 |
| 3.3 | Fixed linear resistors | 55 |
| 3.4 | Variable linear resistors | 57 |
| 3.5 | Non-linear resistors | 61 |

## 4. Capacitors
| | | |
|---|---|---|
| 4.1 | Introduction | 66 |
| 4.2 | Capacitor principles | 66 |
| 4.3 | Electrolytic capacitors | 69 |
| 4.4 | Paper and plastic capacitors | 74 |
| 4.5 | Ceramic capacitors | 76 |
| 4.6 | Mica capacitors | 79 |
| 4.7 | Capacitors for special applications | 80 |
| 4.8 | Capacitor selection | 82 |

## 5. Magnetic components
| | | |
|---|---|---|
| 5.1 | Introduction | 83 |
| 5.2 | Permanent magnets | 83 |
| 5.3 | Transformers | 85 |
| 5.4 | Inductors | 88 |
| 5.5 | Relays | 89 |
| 5.6 | Hall effect devices | 100 |
| 5.7 | Magneto resistors | 104 |

## 6. Peripheral components
| | | |
|---|---|---|
| 6.1 | Introduction | 107 |
| 6.2 | Switches and keyboards | 107 |
| 6.3 | Fuses | 116 |
| 6.4 | Heat sinks | 119 |
| 6.5 | Connectors | 124 |

## 7. Quartz, ceramic, glass and selenium
| | | |
|---|---|---|
| 7.1 | Introduction | 132 |
| 7.2 | Ferroelectricity | 132 |
| 7.3 | Piezoelectricity | 132 |
| 7.4 | Pyroelectricity | 140 |
| 7.5 | Glass | 140 |
| 7.6 | Selenium | 143 |

## 8. Power sources
| | | |
|---|---|---|
| 8.1 | Introduction | 145 |
| 8.2 | Cell characteristics | 145 |
| 8.3 | Primary cells | 146 |
| 8.4 | Secondary cells | 152 |
| 8.5 | Battery selection | 157 |
| 8.6 | Fuel cells | 158 |
| 8.7 | Solar cells | 159 |
| 8.8 | Special cells | 163 |

| | |
|---|---|
| Bibliography | 165 |
| Glossary of acronyms | 166 |
| Glossary of terms | 167 |
| Index | 171 |

# Preface

Electronic components come in many different sizes and shapes, and the first task of an author on the subject is to define the area which he is to cover. In this book I have concentrated on components which are normally mounted on a printed wiring board, or are usually associated with such assemblies. All reference to integrated circuit components has been omitted since this is the topic of a companion volume (*Integrated Circuits*, by F. F. Mazda, Cambridge University Press, 1978).

It would be foolish for me to claim that I have covered all aspects of discrete components used on printed wiring boards. To do so would need a book many times the present size. Instead I have concentrated on the more important aspects of the components, and the less well known facts about them, whilst providing enough of the more elementary data to serve as an introduction for the newcomer.

The book is primarily aimed at electronic engineers in industry, although parts of it should prove useful to postgraduate and undergraduate students. I have minimised the amount of description of physical phenomena and tried instead to describe the construction and characteristics of the different components, and their main selection criteria in electronic circuits.

The first chapter of the book describes discrete semiconductor devices. These vary from diodes through to transistors, thyristors, the gate turn off switch and triacs. Also described in this chapter are the trigger devices usually associated with thyristors and triacs, such as the unijunction transistor.

The terminology and units used to measure optical parameters, which are often confusing to electronic engineers, are described in chapter 2. This is followed by sections on optical sources, detectors and optocouplers. An important aspect of optoelectronic components is displays, and modern developments in this area are described. The chapter concludes with brief introductions to optical communication and holography.

Resistors and capacitors form the subject of the next two chapters. These are probably the most frequently used electronic components, yet there are a bewildering number of different types available commercially. The chapters describe the basic differences between the varieties and provide guides to their selection.

Magnetic components covers a wide spectrum of devices. Chapter 5 describes the characteristics and construction of permanent magnets, transformers, inductors and relays. Included in this chapter are semiconductor devices which operate from a magnetic field, such as Hall effect devices and magneto resistors.

Chapter 6 looks at components which are generally auxiliary to the main printed wiring board circuit. Examples of these are switches, fuses, heat sinks and connectors. The components described in chapter 7 are based on the materials glass, quartz, ceramic and selenium. The book concludes with a chapter on power sources covering the conventional primary and secondary cells, and the more novel fuel and solar cells.

I am grateful to many friends and colleagues for useful discussions during the preparation of this book, and especially to Dr H. Ahmed of the University Engineering Department, Cambridge University, for reading through the draft of this book and for his invaluable comments.

F. F. Mazda

Sawbridgeworth 1981

# 1. Discrete semiconductors

## 1.1 Introduction

Discrete semiconductors fall into four main groups: diodes, transistors, thyristors and triacs. Each of these groups consists of several different types of components designed for specific applications. This chapter first describes the techniques used in the manufacture of semiconductor devices and then describes the construction and characteristics of the different devices.

## 1.2 Semiconductor production techniques

The prime requirement of a semiconductor manufacturing *process** is the ability to prepare a material which has the minimum number of defects in its crystal structure, and which is doped with the required amount of *impurities*. To do this, the raw semiconductor is first refined. Then it is grown as a bulk single crystal to which various impurities are added.

### 1.2.1 *Semiconductor crystal preparation*

Several techniques exist for growing bulk crystals. In most of these the crystal is grown from a melt, since, as the crystal is drawn off, impurities tend to remain in the melt which can eventually be discarded. Fig. 1.1*a* shows one such system, the zone levelling method. A silica crucible containing silicon is slowly moved along the tube such that the molten zone under the heating coils moves down the crystal, carrying impurities with it. After several passes, the end of the crystal where impurities have been concentrated can be cut off and discarded. An inert atmosphere is introduced into the tube and the crucible material must be such that it does not contaminate the silicon. Gallium arsenide may also be grown by this technique but it is much more easily contaminated. In order to lessen contamination, a sealed silica crucible which also contains arsenic gas under high pressure is used. This prevents the gallium arsenide from decomposing and thus reduces contamination.

The zone levelling method is relatively simple, but since the crystal cools in contact with the crucible the crystal quality tends to be poor and the risk of contamination is high. Fig. 1.1*b* shows an alternative system, called Czochralski growing, which avoids contact between the crystal and crucible. A tiny seed crystal of silicon is lowered into the molten silicon and slowly rotated and withdrawn. The silicon from the melt settles on the seed crystal and cools, and a bar of pure silicon is gradually formed.

Figs. 1.1*c* and *d* show the floating zone refining and the pedestal pulling methods in which neither the melt nor the crystal are in contact with the crucible so avoiding the risk of contamination. In the method of floating zone refining the heating coils slowly move down the crystal rod, and in the pedestal pulling method the seed crystal is dipped into the melt and then slowly withdrawn. The melt in both cases is held in place by surface tension. This means that the melt cannot be very large in size.

### 1.2.2 *Epitaxial growth*

In all the methods shown in fig. 1.1 a measured amount of p or n impurity can be added to the melt to give a doped silicon rod. This rod may then be cut into slices with a diamond impregnated saw. The top surface of the slice, which is damaged by the saw, is removed by etching, and the slice is polished with fine diamond powder. At this stage in its production the slice is usually referred to as the bulk semiconductor. It is now ready for subsequent processing steps. One of these consists of growing a thin layer of material which forms a molecular extension of the main material on one side of the slice. This layer, called an *epitaxial layer,* is generally purer

---

* The first occurrence of words that are included in the glossary of terms (p.167) are given in italic.

## 2  Discrete semiconductors

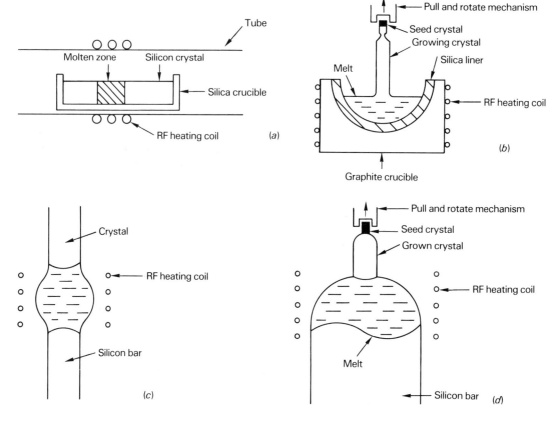

Fig. 1.1. Semiconductor bulk crystal purification and growth systems; (a) zone levelling, (b) Czochralski growing, (c) floating zone refining, (d) pedestal pulling.

than the bulk material and can have its impurity concentration more precisely controlled. The method of epitaxy formation consists of heating the semiconductor slices in a furnace through which a reactive gas is passing, as shown in fig. 1.2. When silicon slices are being processed the gas used is hydrogen which has been bubbled through silicon tetrachloride. In the presence of the heated silicon slices the silicon from the gas is deposited on the slice leaving hydrochloric acid gas, which is removed. The epitaxy layer can be doped by a p or n impurity by first bubbling the hydrogen through the corresponding impurity solution. When gallium arsenide slices are being processed a modified system is used; the hydrogen gas is saturated with arsenic by bubbling it through $AsCl_3$ before being passed over the gallium arsenide. The $AsCl_3$ produces a gas phase mixture, and from this the gallium arsenide epitaxy is deposited onto the slices.

### 1.2.3 Diffusion

An alternative method for introducing impurities into the bulk semiconductor surface is *diffusion*. A furnace similar to that shown in fig. 1.2 but at the much higher temperatures at which the

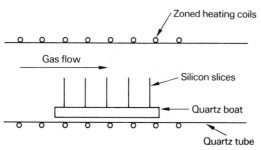

Fig. 1.2. Epitaxy furnace.

impurity atoms are very mobile, is used. The impurities are passed over the heated semiconductor substrate (i.e. slice) either by bubbling nitrogen through a liquid form of the impurity or by placing the substrate and impurity crystals in a sealed boat in the heated tube. In the latter case, the boat must be broken to remove the slices at the completion of the diffusion process. Although diffusion is slower than epitaxial growth it allows much finer control of the impurity concentration and depth.

### 1.2.4 *Ion implantation*

Impurities can also be introduced into the bulk substrate by accelerating impurity ions in an electric field and then bombarding them onto the substrate. This is known as *ion implantation*. The depth to which the ions penetrate depends on the type of ion used, the composition of the substrate material and the strength of the electric field. Ion implantation causes damage to the surface crystal structure of the substrate so it is normally followed by annealing. The annealing temperatures must be carefully controlled to prevent excessive diffusion of the implanted impurities.

### 1.2.5 *Photolithography*

In the preparation of semiconductor devices it is often required to control the location at which the impurities are introduced. This is done by coating the silicon surface with a material such as silicon dioxide and then selectively removing it from the places where the impurities are to be introduced. The silicon dioxide (also called silica or, simply, *oxide*) is grown on the silicon surface by heating a silicon slice in a tube such as that shown in fig. 1.2, and then passing dry or wet oxygen or steam over its surface. To remove this oxide layer selectively photolithographic techniques similar to those used in printed circuit board manufacture are used. The oxide surface is coated with a photoresist and then covered with a mask, usually a glass plate with the areas to be removed covered in opaque emulsion. The plate is then exposed to ultraviolet radiation which will harden the exposed areas. The unexposed photoresist is now removed with a developer and the oxide below it is then etched with an etchant which attacks the oxide but not silicon. The hardened photoresist areas are then removed in a separate process to leave the silicon slice with the oxide layer removed in selected areas. Diffusion introduces impurities into the oxide layers much more slowly than into the exposed silicon areas, and so can be used to dope the desired areas.

### 1.2.6 *Vacuum deposition*

The final semiconductor production process to be described here is known as vacuum deposition. It is used for putting layers of metal or other materials onto the semiconductor surface. Two techniques exist for this. In the first system, the semiconductor is held face down in the top half of an activated bell jar, and the material to be deposited is located at the bottom of the jar and heated until it vaporises and settles as a thin layer onto the semiconductor surface. The second deposition technique is called sputtering. A bell jar is again used but now it is filled with an inert gas such as argon. The semiconductor, again face down in the top half of the jar, is connected to the positive terminal of a high voltage source and the material to be deposited is placed at the bottom of the jar and connected to the negative terminal. Under the effect of this voltage the argon is ionised and positive ions bombard the cathode causing it to sputter and emit material which settles on the semiconductor surface (anode).

## 1.3 Diodes

The symbol for a diode is shown in fig. 1.3 and its d.c. *characteristic* in fig. 1.4. In the forward direction the current increases rapidly through the device once its internal potential $V_1$ is overcome. In the reverse direction there is much less current flow, called the leakage current. For large values of reverse voltage ($V_2$) the diode will break down into avalanche conduction. The current rapidly increases and since the voltage across the device is still $V_2$, which in a conventional diode can be several hundred times the value of $V_1$, the power dissipation in the avalanche mode may be very large and this will destroy the device.

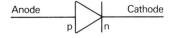

Fig. 1.3. Diode symbol.

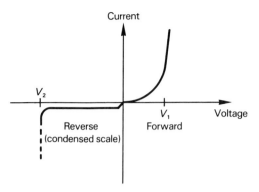

Fig. 1.4. Diode characteristics.

### 1.3.1 Diode ratings

*Data sheets* specify semiconductors by two sets of parameters, *ratings* and *characteristics*. The ratings define the operating limits of the diode which, if exceeded, could damage it. The characteristics define its performance under certain conditions. The following are some of the ratings of a diode.

(i) Maximum reverse voltage. This is usually specified in three ways: (*a*) the peak working voltage, which is the normal operating voltage of the device; (*b*) the peak repetitive voltage, usually higher than the peak working voltage, but which the diode will stand for short periods; (*c*) the peak non-repetitive voltage. This must occur only infrequently during the life of the device since it causes the highest power dissipation, and therefore strain, to the silicon dice.

(ii) The maximum current rating. This is also defined by three parameters, i.e. peak working current, peak repetitive current and peak non-repetitive *surge* current. The working current may be specified as a direct current or as an average value.

(iii) Peak power dissipation.

(iv) Maximum temperature at which the diode junction can be operated and stored. This maximum value includes the hottest and coldest temperatures.

Some of the characteristics of a diode are:

(i) Forward voltage drop. This varies with forward current as shown in fig. 1.4 so it is normally specified at a given current, or by a graph. The average forward voltage drop is usually specified as an average over a full cycle at a stated

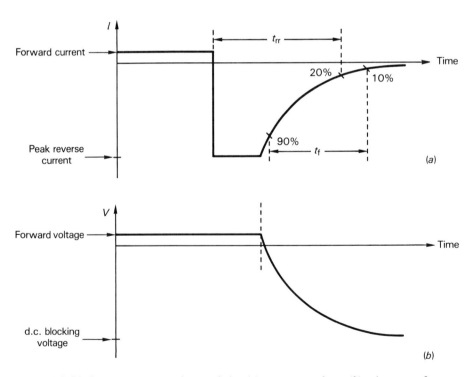

Fig. 1.5. Diode reverse recovery characteristics; (*a*) current waveform, (*b*) voltage waveform.

mounted between copper discs to facilitate cooling. Fig. 1.6c shows a stud arrangement which enables the package to be screwed down to a heat sink. The silicon chip is usually double diffused having p and n impurities diffused into opposite sides of the wafer. For very large power ratings the press pak package is used since it enables the diode to be sandwiched between two heat sinks.

### 1.3.3 Schottky diodes

In the Schottky barrier diode, shown in fig. 1.7a, rectification takes place at the metal–semiconductor interface and this gives it a low reverse recovery time characteristic. The n layer is only lightly doped and this causes a potential barrier to be set up at the interface between the metal and the silicon. For conduction to occur electrons must jump this barrier by thermal excitation, and this has led to the device also being called a hot carrier diode. The disadvantage of the Schottky diode is that it has a higher reverse leakage current than conventional devices and therefore a lower reverse voltage rating. The anode metal is usually molybdenum with a layer of gold on top for circuit connections.

The disadvantage of the simple structure shown in fig. 1.7a is that a high electric field occurs at the edges of the diode under reverse *bias* conditions which gives a lower reverse voltage rating. This can be overcome by the mesa arrangement (fig. 1.7b) or by the hybrid technique in which p diffusions are used to prevent the high fields under reverse bias conditions (fig. 1.7c).

### 1.3.4 Varactor diodes

The capacitive effect of a pn junction is used for applications such as electronic tuning, harmonic generation and parametric amplification. Such a device is known as a varactor diode and its typical doping profile is shown in fig. 1.8a. Two types of construction are used for this diode, planar diffused and mesa.

The planar diffused device has a high resistance since the current flow is sideways through the n epitaxy layer, which is made of high resistivity material in order to give devices with high voltage ratings. This disadvantage is overcome in the mesa arrangement.

The varactor diode characteristic is shown in fig. 1.8d. The non-linear capacitance-voltage curve results in an output, from the diode, which is rich in harmonics. This can be filtered to generate any required high frequency, so it acts as a frequency multiplier. Both silicon and gallium arsenide are used to make varactor diodes. Gallium arsenide has a lower resistance since it has a higher carrier mobility, and it is best suited to high frequency low noise applications. Silicon has a longer *carrier* lifetime so it can give rise to diffusion capacitance in the forward direction at high frequencies when no appreciable recombination occurs. It is therefore used for appli-

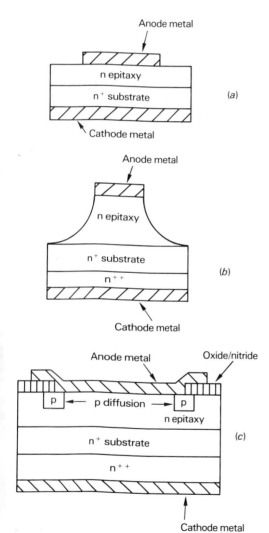

Fig. 1.7. Schottky barrier diode; (a) elementary structure, (b) mesa arrangement, (c) hybrid structure.

frequency, average forward current and case temperature.

(ii) Reverse leakage current. This is specified at a defined reverse voltage or as an average value over one cycle.

(iii) Forward power dissipation. This is the product of forward volt drop and current and is given as a curve.

(iv) Junction to case thermal resistance. This characteristic, defined in units of centigrade degrees per watt, indicates the difference in temperature between junction and case when the diode is dissipating power. It is required for designing diode cooling systems, as described in chapter 6.

(v) Diode capacitance. This is responsible for what is known as its reverse recovery effect, as shown in fig. 1.5. When the diode is conducting in the forward direction an excess of charge is stored in the device. If the voltage is now rapidly reversed the charge at the extremities of the silicon *die* will be swept out and give rise to a large reverse current. Carriers near the junction, however, disappear due to *recombination* with each other, and only when this process has been completed does the diode recover its voltage blocking capability. The diode *recovery time* can be reduced by using fast recovery diodes, such as the Schottky diode described in section 1.3.3, or by introducing impurities, like gold, into the junction, which trap the free carriers. Gold doping has the disadvantage of increasing the diode's forward volt drop and reverse leakage current. It can be seen from fig. 1.5 that the reverse current flow must not stop abruptly as this will give rise to large voltage spikes in any associated inductive circuits. The ratio of $t_f/t_r$ must be as large as possible and devices which have a relatively large value of this ratio are said to be 'soft'.

### 1.3.2 *Diode construction*

Fig. 1.6 illustrates the main types of diode. In the alloyed structure (fig. 1.6a) aluminium used as the p *dopant* to the n type silicon. The silicon is in direct contact with the cathode whereas a spring contact is used to connect aluminium to the anode lead. In the planar fused structure (fig. 1.6b) both leads butt up the silicon chip giving a more robust construction. For power applications the silicon chip

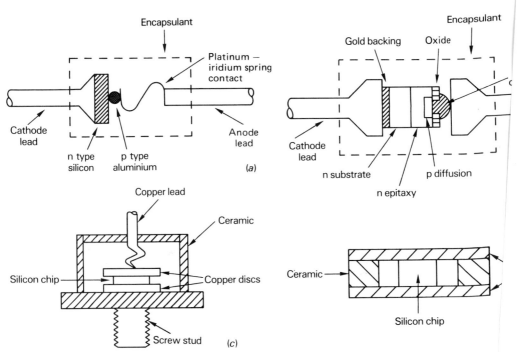

Fig. 1.6. Diode configurations; (a) alloyed, (b) planar diffused, (c) stud package, (d) press pak pack

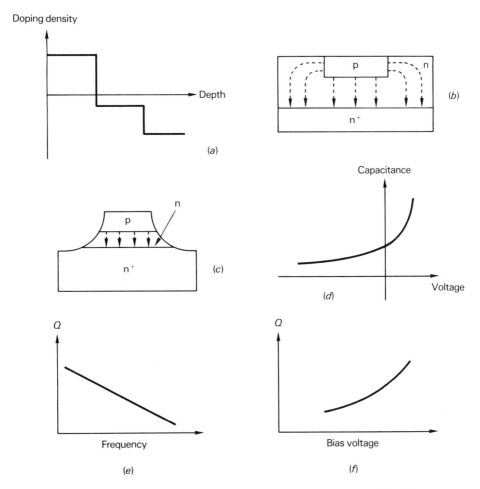

Fig. 1.8. Varactor diode; (a) doping profile, (b) planar construction, (c) mesa construction, (d) capacitance characteristics, (e) variation of $Q$ with frequency, (f) variation of $Q$ with bias voltage.

cations needing a wider frequency band. The figure of merit for a varactor diode is given by its quality or $Q$ factor which is defined by

$$Q = 1/2\pi fCR, \qquad (1.1)$$

where $f$ is the frequency, $C$ is a combination of the diode junction capacitance and the stray capacitance due to the case, and $R$ is the series resistance. Figs. 1.8e and f illustrate typical curves of the variation of the $Q$ factor in a varactor diode.

### 1.3.5 Step-recovery diodes

A type of varactor diode which is especially efficient for frequency multiplier applications is the step-recovery diode. Its construction and doping profile are shown in fig. 1.9. The intrinsic region is swept clear of charge carriers and for low reverse voltages this results in the device having a small but constant capacitance. For forward voltages the diffusion capacitance is large and results in a diode with negligible impedance. The operation of the diode can be explained with reference to the current and voltage waveforms shown in fig. 1.9d. During the forward voltage, minority carriers are injected into the junction. If the voltage is now reversed in a time period which is short compared to the carrier lifetime, then the minority charge will be recovered as a large reverse current which will stop abruptly giving rise to a distorted current waveform which is rich in harmonics.

# 8  Discrete semiconductors

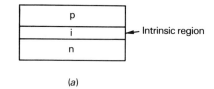

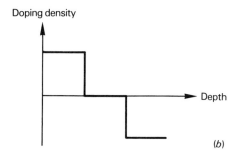

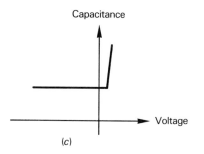

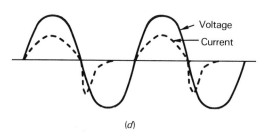

Fig. 1.9. Step-recovery diode; (a) construction, (b) doping profile, (c) capacitance characteristic, (d) voltage–current curve.

### 1.3.6 Negative resistance diodes

Fig. 1.10 shows the characteristic of a class of diodes which exhibit a *negative resistance,* and are primarily used at microwave frequencies. Due to their negative resistance characteristic they can be used in oscillators, amplifiers or switches.

One of the devices which exhibit the characteristic shown in fig. 1.10 is the tunnel diode,

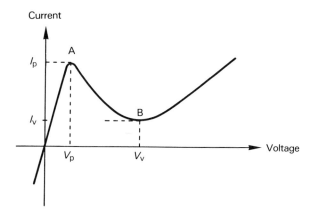

Fig. 1.10. Negative resistance diode characteristic.

which has a doping density some thousand times greater than in a conventional diode. The high doping density results in a very narrow depletion region and so transfer of holes and electrons occurs predominantly by tunnelling. When the diode is reverse biased a large number of empty states are formed on the n side so that electrons tunnel across the junction resulting in a large current flow. In the forward direction the current rapidly increases due to tunnelling. This soon results in a shift in the electron distribution on both sides of the junction, and a fall in the tunnelling current between typical points A and B shown in the figure. After point B, the current again increases with voltage as in a conventional diode. The maximum power output from the tunnel diode is proportional to the product of the voltage and current variations, i.e. to $(V_v - V_p)(I_p - I_v)$. The voltage variation can be increased by reducing the doping level, but this reduces current variation due to a reduction in $I_p$, so that the product of the voltage and current variations is not substantially increased. Therefore, the power output of a tunnel diode is relatively small although it produces very low noise at high frequencies.

Fig. 1.11 shows some other diodes which exhibit a negative resistance characteristic, and which have been used for high frequency applications. The transferred electron device is made from gallium arsenide or indium phosphide, and not from silicon. It operates in any of three modes, transit time (also called Gunn effect), quenched domain, and limited space charge accumulation (LSA). These devices are tunable

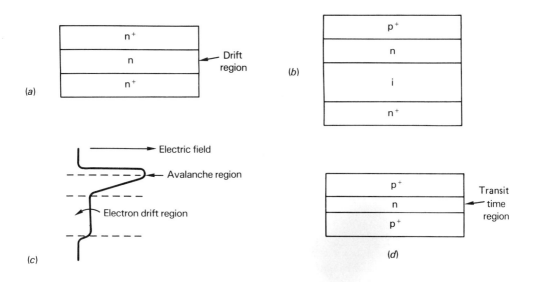

Fig. 1.11. Some negative resistance diodes; (a) transferred electron diode, (b) IMPATT diode, (c) IMPATT doping profile, (d) BARITT diode.

over a broad frequency range and can produce high power outputs. The LSA mode can generate the highest power since the whole diode characteristic is in its negative resistance region.

The impact avalanche transit time (IMPATT) diode can be made from gallium arsenide or silicon. It combines impact avalanche breakdown and charge carrier transit time effects. When a voltage is applied across the diode which exceeds its junction breakdown value, hole-electron pairs are formed due to avalanching. These carriers pass through the drift region and generate a power output. The length of the electron drift region can be chosen such that the terminal current is in antiphase with the applied voltage, so producing maximum negative resistance. In a conventional IMPATT diode avalanching occurs over most of the depletion region. A Read diode is a development of the IMPATT diode and uses special processing techniques to confine avalanching to a narrow region. An IMPATT diode can give a greater power output than a transferred electron device but it also needs a higher operating voltage and produces more noise.

A variation of the IMPATT diode is the trapped plasma avalanche transit time (TRAPATT) diode. It has an identical structure to the IMPATT diode but it operates in a different mode which requires higher current levels and lower frequencies. The TRAPATT diode effect occurs in lightly doped layers of $n^+pp^+$ or $p^+nn^+$ diodes. The diode itself is usually made from silicon. It can produce higher pulsed power outputs than IMPATT diodes.

The barrier injection transit time (BARITT) diode has two junctions. One of these is forward biased and injects carriers into the transit time region, whereas the other junction is reverse biased and collects these carriers. The time taken by the carriers to cross the intermediate region gives rise to a phase difference between the current and applied voltage, and this produces a negative resistance region. The BARITT diode has a low power output and operates at relatively low frequencies. However it has very good sensitivity and is used in consumer radar applications.

Fig. 1.12 summarises the properties of three materials which have been used to make the diodes shown in fig. 1.11. A high electron mobility results in low loss, high *gain* and a large power output. A large electron drift velocity gives a short electron transit time and is therefore required for high frequency operation. Thermal conductivity determines the rate of heat removal. The higher the value the greater the power dissipation and reliability of the device at high power levels.

## 10  Discrete semiconductors

| Parameter | Silicon | Gallium arsenide | Indium phosphide |
|---|---|---|---|
| Electron mobility | 3 | 1 | 2 |
| Electron drift velocity | 3 | 2 | 1 |
| Thermal conductivity | 1 | 3 | 2 |

Fig. 1.12. Comparison of semiconductor materials; 1 = highest, 3 = lowest.

### 1.3.7 Voltage reference diodes

A reverse biased pn junction will break down due to avalanching if the voltage is made large enough. It is possible, by controlling the doping concentration, to construct a pn junction that breaks down at a much lower voltage. This is called the zener effect. The breakdown curve is much 'softer' than that of an avalanche mode and is shown in fig. 1.13b. The device is called a voltage reference or zener diode and its symbol is shown in fig. 1.13a. In the forward direction it behaves like a conventional diode.

The zener voltage of a voltage reference diode can be varied by controlling several parameters, one of which is the material resistivity. The maximum current rating of the diode is set by this voltage and by the power rating. The reverse leakage current of the diode is usually specified at 80% to 90% of its zener voltage. Since a zener diode is normally used to produce a stable d.c. voltage it should have a zener voltage which does not vary appreciably with zener current or temperature. The change in zener voltage for a change in zener current is called the *dynamic resistance* of the diode and this should be as small as possible. Fig. 1.13c shows that the dynamic resistance is dependent on the current and the zener voltage of the diode. It is minimum at a voltage of between 5.5 and 6.5 V. Fig. 1.13d shows the variation of zener voltage with temperature and the rated zener voltage of the diode. A diode having a zener voltage between 5 and 6 V does not show any appreciable change of voltage with temperature.

Several techniques are used in the manufacture of zener diodes. The simple diffused structure of fig. 1.14a has the disadvantage that the edges of the pn junction are exposed to contamination. In the *passivated* structure of fig. 1.14b the edge of the pn junction, which is situated at the surface of the silicon, is covered by a layer of silicon dioxide and is therefore

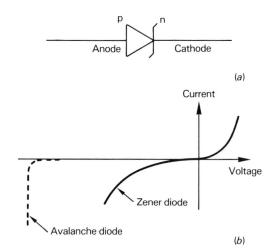
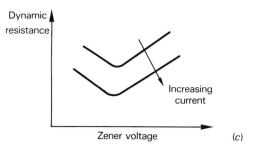
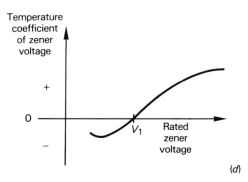

Fig. 1.13. Voltage reference diode; (a) symbol, (b) d.c. characteristic, (c) dynamic resistance, (d) temperature coefficient of zener voltage.

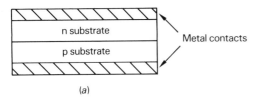

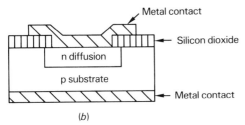

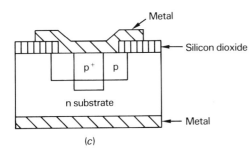

Fig. 1.14. Voltage reference diode construction; (*a*) diffused, (*b*) diffused and passivated, (*c*) alloy diffused.

protected from contamination. Similarly, all junctions are protected in the alloy diffused structure (fig. 1.14*c*). This structure exhibits better zener characteristics at lower zener voltages. The diffused and passivated structure is used at higher zener voltages. The temperature compensated voltage reference diode or stabistor is a modification of the conventional zener diode. It consists of a series arrangement of zener diodes and ordinary diodes. Temperature compensation is achieved by matching the equal and opposite *temperature coefficients* of the zener and ordinary diodes. In the forward direction the breakdown voltage of the stabistor is equal to the breakdown voltage of the series diodes and can therefore be large. In the reverse direction the voltage is the zener voltage plus the diode voltage drops. Since these vary with the magnitude of the zener current, the temperature coefficient of the zener voltage is also dependent on the current.

## 1.4 Transistors

### 1.4.1 *Bipolar transistors*

*Bipolar* transistors may be npn or pnp. The base current of a bipolar transistor is much less than its collector current and the ratio $I_C/I_B$ is known as the static common emitter gain ($h_{FE}$ or $\beta$) of the transistor. $h_{FE}$ is dependent on collector current and falls at large current values, as shown in fig. 1.15*a*. Other parameters which are of interest when using transistors are ratings such as the maximum collector-emitter, base-emitter and collector-base voltages, the maximum collector and base currents, and the power dissipation. Apart from the static gain, the characteristics include the leakage (collector cut-off current) the saturation voltage drop across the collector-emitter at a specified collector current and capacitances of the input and output. Fig. 1.15*b* shows a typical input characteristic. It is primarily determined by the forward biased base-emitter diode junction. The higher the value of $V_{CE}$ the wider the collector-base depletion layer becomes. This gives a thinner effective base conducting region. The recombination in the base area is reduced so that for a given $I_B$ a higher value of $V_{BE}$ is now required. The effect of $V_{CE}$ on the input characteristic is generally not very significant.

The bipolar transistor output characteristic is shown in fig. 1.15*c*. For very low values of $V_{CE}$ the collector does not gather up electrons which pass through the base region. This process becomes more efficient as $V_{CE}$ increases giving a rapid rise in $I_C$. After the 'knee' of the curve has passed, most of the carriers generated at the base-emitter junction are gathered up by the collector so that increasing $V_{CE}$ does not appreciably effect $I_C$. This is known as the saturation region of the characteristic. The larger the value of $I_B$ the higher the collector current at which saturation occurs. Therefore a bipolar transistor is essentially a current operated device. At very large values of $V_{CE}$ the carriers attain sufficient energy to cause avalanche breakdown.

Figs. 1.15*d* and *e* show the switching waveforms of a transistor. Assuming that the base voltage changes at $t_1$ there is a delay until $t_2$ before the transistor starts to turn on. Thereafter the output changes relatively slowly while stray circuit capacitances are discharged, until at $t_3$ the transistor is fully on. During switch off there is a delay called the storage time before the transistor begins to respond and then a further fall time, determined by stray capacitances, before the device attains its fully off state. *Rise time* and *fall time* are important parameters for transistors which are to be used in high frequency or switching applications.

### 1.4.2 Unipolar transistors

*Unipolar* transistors, unlike bipolars, have a conduction mode which depends on only one carrier, which may be holes or electrons according to the type of device. Fig. 1.16a shows the operation of a junction unipolar transistor or junction field effect transistor (JFET) as it is often called. In one type of structure the *gate* region is formed by two p diffusions into an n type silicon substrate. The two pn junctions exhibit depletion regions which extend almost entirely into the n region since it is doped much less heavily than the p areas. With the battery arrangement shown the gate-source and gate-drain regions are reverse biased. Since the gate-drain has a higher reverse bias, the depletion region extends further. If $V_{GS}$ is kept constant and $V_{DS}$ is increased, then the depletion regions will extend until they almost meet. This is known as the pinch-off state and a further increase in $V_{DS}$ does not affect the drain current $I_D$. The more negative $V_{GS}$ is made the lower the value of $V_{DS}$ and therefore $I_D$ at which pinch-off occurs.

Fig. 1.16b shows the structure of a JFET. It is *n channel* since electrons are the charge carriers between source and drain. A *p channel* JFET would have a p substrate and a n diffused gate, and holes would now be the carriers. Symbols for p and n channel devices are shown in figs. 1.16c and d, the arrow pointing from the p to the n regions.

In the JFET, current flows through the device in the absence of a gate voltage, and this is known as *depletion mode* operation. An alternative unipolar transistor, called the metal oxide semiconductor FET (MOSFET) or insulated gate FET (IGFET), is shown in fig. 1.17a. This usually operates in an *enhancement mode* since negligible current flows between source and drain at zero gate bias. However, when the gate is made positive it attracts electrons so that a channel is formed under it between source and drain giving rise to a current. It is possible to dope the channel region of a MOSFET such that it operates in a depletion mode, to give the output characteristic shown in fig. 1.17b. The transfer curves for n channel and p channel devices are also given in figs. 1.17c and d, a p channel

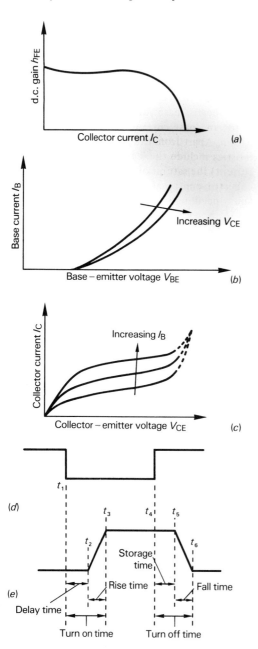

Fig. 1.15. Transistor characteristics; (a) d.c. gain, (b) input, (c) output, (d) switching base voltage, (e) switching collector voltage.

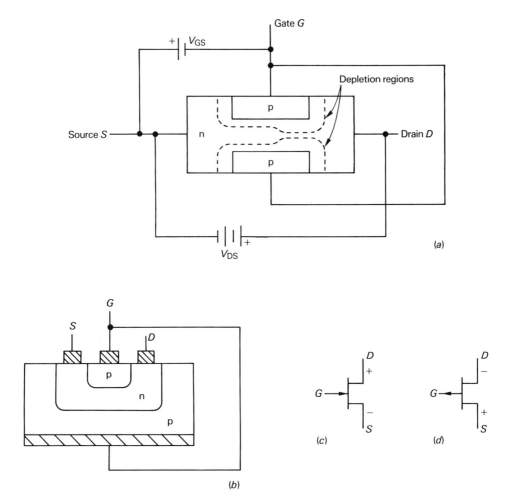

Fig. 1.16. Junction field effect transistor (JFET); (a) representation of n channel device; (b) practical arrangement of n channel device, (c) n channel symbol, (d) p channel symbol.

MOSFET usually operating in an enhancement mode only. The symbols for a MOSFET show the insulated gate arrangement since it is separated from the silicon substrate by the silicon dioxide layer. The arrow head indicates whether the substrate is p or n type. A continuous line between source and drain means that current flows even in the absence of a gate potential, i.e. it is depletion mode, and a broken line is used for enhancement mode transistors.

The MOSFET has a much higher input impedance than a JFET and this is independent of the polarity of the gate voltage. It also has a lower leakage current and because of the simpler MOSFET structure it can have more devices integrated onto a single silicon chip than a JFET. A JFET on the other hand has a higher breakdown voltage than a MOSFET, it shows better stability, is more robust and it has a lower ON resistance. It is only available as a depletion mode device.

The main difference between a unipolar and bipolar transistor is that the FET is voltage operated and it is primarily a resistive element, so it does not exhibit an offset voltage. A FET can also be run symmetrically, with the junctions of the source and drain reversed, and this is useful for several switching applications.

14  *Discrete semiconductors*

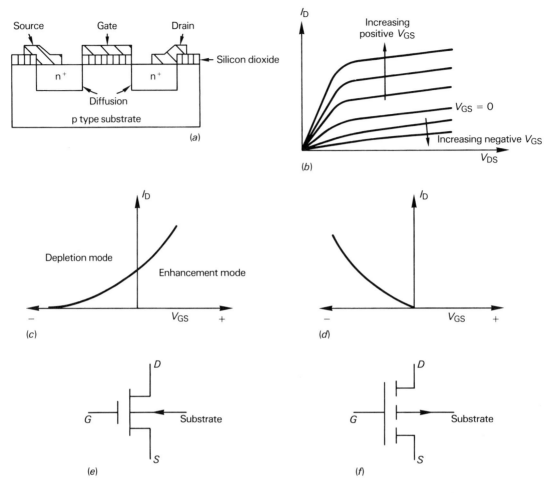

Fig. 1.17. Metal oxide semiconductor field effect transistor (MOSFET); (a) construction, (b) $I_D/V_{DS}$ characteristic, (c) transfer curve for n channel device, (d) transfer curve for p channel device, (e) symbol for n channel depletion mode, (f) symbol for p channel enhancement mode.

### 1.4.3 *Transistors for power applications*

Many different transistor structures are commercially available to meet a range of applications. Fig. 1.18 shows different types of power bipolar devices. The hometaxial transistor is rugged, cheap, has good overall voltage ratings, but has a long switching time. The epi-base device has a low voltage rating. The voltage rating can be increased by adding the high resistive collector epitaxy which causes the collector voltage to be shared by the base and collector epitaxy layers. However the voltage rating is still less than that of a hometaxial transistor. The epi-base device uses shallow emitter regions and a narrow base epitaxy and this gives it a higher speed and current handling capability, for the same emitter area, than a hometaxial transistor.

In the planar epitaxial structure all the junctions are protected by an oxide layer and this gives it a very low leakage current. Narrow base widths are also possible due to the diffusion process, and this enables the transistors to be made with higher speeds and lower saturation voltages than the hometaxial and epi-base devices. The planar epitaxial transistor is not very resistant to the effect of second breakdown, as explained in the next paragraph. The triple diffused structure shows some improvements over

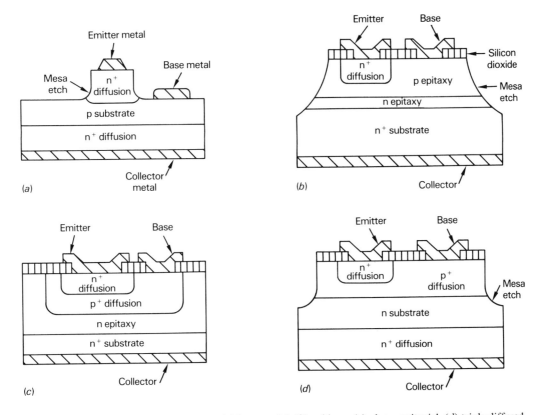

Fig. 1.18. Power bipolar transistor structures; (a) hometaxial, (b) epi-base, (c) planar epitaxial, (d) triple diffused.

planar epitaxial regarding second breakdown, but it is still poor. It has speeds and voltage drops which are similar to planar epitaxial structures but it is more expensive and shows higher leakage.

An important consideration in power transistor design is the second breakdown effect. It can be explained by reference to the curve of fig. 1.19a. For a given base current the collector current $I_C$ will increase as collector-emitter voltage $V_{CE}$ increases. After a point A on the characteristic the device will go into saturation and the current will remain substantially constant until point B when avalanche breakdown, or first breakdown, occurs. This causes a rapid rise in current until point C when a second breakdown effect develops. This results in rapid local heating of the silicon die, the collector-emitter voltage collapses, and the current escalates destroying the transistor. There is a series of curves for different base currents and these give rise to individual second breakdown points, which all lie on a locus, as shown. As the duty cycle of the transistor decreases it runs cooler so that it can work on a wider second breakdown locus.

Power transistor parameters are similar to those of conventional devices except that they must be operated in a mode such that second breakdown is avoided. This is done using *safe operating area* (SOA) curves of the type shown in fig. 1.19b. Although these curves are for a device rated at a peak current of $I_M$ and a voltage of $V_M$ the transistor cannot be run at this current and voltage simultaneously. For low values of $V_{CE}$ the current can increase to $I_M$ where it is limited by the current carrying capability of the bonding wire and the metallisation tracks used on the silicon. As $V_{CE}$ increases so also does the power dissipation so that eventually $I_C$ will need to be decreased. For large values of $V_{CE}$ the value of $I_C$ is reduced still further in order to prevent the occurrence of second breakdown effects. The SOA of the transistor increases as

16  *Discrete semiconductors*

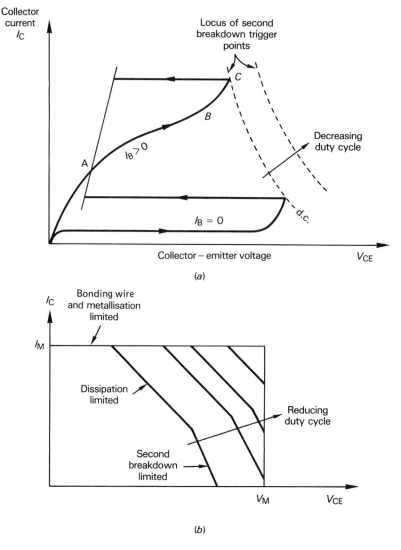

Fig. 1.19. Safe operating area characteristics of power transistors; (*a*) second breakdown characteristic, (*b*) safe operating area curves (SOA).

the duty cycle reduces since both the dissipation and second breakdown effects are now lower.

The disadvantage of a power bipolar transistor is that it needs large base currents since it often has poor gain. Darlington transistors may now be used, which basically consist of two transistors, the emitter of the first transistor acting as an amplifier and feeding the base of the second. But they are expensive and have a larger voltage drop. The alternative is to use a power FET, which is a voltage operated device. It has the added advantage of being immune to second breakdown and thermal runaway effects although, since it has a relatively large ON resistance, it is usually available only at lower current ratings than a bipolar transistor. Fig. 1.20 shows a unipolar structure called a V-groove metal oxide semiconductor (VMOS) which is capable of high power ratings. The high current of the device is partly due to the short channel spacing, the presence of two current paths and to the position of the drain on the back face of the silicon chip. The high resistance n region between drain and source allows the channel-drain depletion to spread into the drain so that peak electric fields are reduced and the transistor breakdown voltage rating is increased.

### 1.4.4 Transistors for high frequency applications

Special design techniques are required for transistors which are to be operated at very high frequencies. In bipolar devices operating at high

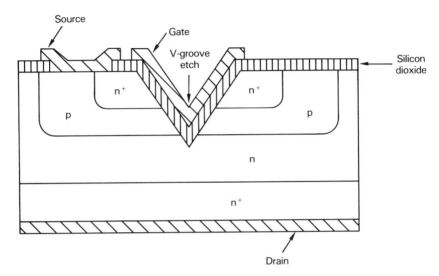

Fig. 1.20. V-groove metal oxide semiconductor (VMOS) power field effect transistor.

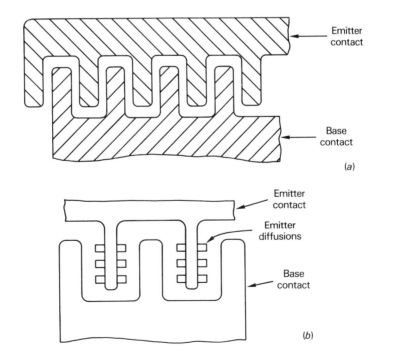

Fig. 1.21. High frequency transistor geometries; (a) interdigitated, (b) overlay.

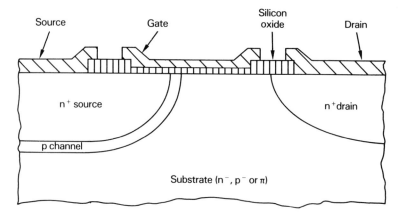

Fig. 1.22. Double diffused metal oxide semiconductor (DMOS) transistor structure.

frequencies the current is forced out towards the edges so that the ratio of emitter periphery to area is large. Two different techniques are used to accommodate this as shown in fig. 1.21. In the interdigitated arrangement the base and emitter regions are interleaved and formed on the silicon die. The diffusions for both these lie under the metal contact areas shown. In the overlay technique the emitter metal contact lies over the base instead of adjacent to it. The emitter diffusions are segmented and spread out from the emitter metal contact, whereas the base diffusion lies below and along the length of the base metal contact.

In unipolar transistors the speed of operation is often determined by the length of the channel. For high frequency operation this should be as short as possible, but it is limited in practice by the photolithographic process. Fig. 1.22 shows a double diffused MOS transistor (DMOS) in which p and n diffusions are used and the channel length is determined by their different diffusion depths, which can be very closely controlled.

Alternative materials such as gallium arsenide make faster devices than silicon and considerable research has gone into developing transistors using these alternative materials. However, although p diffusions in gallium arsenide are fairly well characterised, the n diffusions are limited to very shallow regions. It is possible to use Schottky gates for FET devices, giving the added advantage of negligible charge storage due to minority carriers. Fig. 1.23 shows the construction of such a transistor which is called a MESFET. The higher electron speed and mobility of gallium arsenide over silicon means that it can produce a larger power gain and power output for the same frequency and output impedance.

An important consideration in a high frequency transistor relates to its package. This needs to have a low stray capacitance and lead inductance. Cooling is also important at gigahertz frequencies and *heat pipe* techniques discussed in chapter 6 are often used within the package. Fig. 1.24 shows one type of high frequency package. Twin emitter leads are used which gives symmetry of board layout when devices are combined for greater power. The leads are low inductance strip lines which can interface to microstrip lines used in UHF-VHF equipment. Beryllium oxide is used as the chip insulator since, it has good thermal conductivity, and the chip is located onto a copper stud which is bolted to a heat sink.

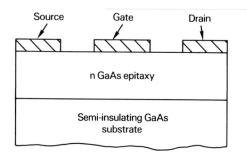

Fig. 1.23. Gallium arsenide field effect transistor.

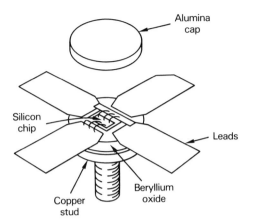

Fig. 1.24. High frequency transistor package.

## 1.5 The thyristor

The thyristor has four semiconductor layers and up to four terminals, as shown in fig. 1.25a. The operation of the thyristor is best understood by considering it to be made up of two transistors as shown in figs. 1.25b and c. With no signal at terminals $G_1$ and $G_2$ both transistors are normally off and the device will block voltage between terminals $A$ and $K$. As the anode to cathode voltage increases so also does the leakage current, which flows as a base current. When this reaches a sufficiently large value it causes both transistors to turn on. Once conducting, the transistor pair will stay on as long as the current through them does not fall below a minimum value called the holding current. If the current falls below the holding current the base currents through the transistors are not adequate to keep them on and they turn off, and once again the applied voltage is blocked. Although the thyristor can be triggered by a signal at gate $G_1$ or $G_2$, in practice it requires a much larger drive at $G_2$, so that in a conventional device only terminal $G_1$ is connected. The symbol for the thyristor is shown in fig. 1.25d. It is a *rectifier* since it can only conduct current from anode to cathode. The device turns on only when a sufficiently large current is passed into the gate terminal $G$.

### 1.5.1 Thyristor characteristics and ratings

Fig. 1.26a shows the static thyristor characteristics. In the reverse direction it behaves like a diode, blocking voltage until the reverse voltage $V_2$ is reached when avalanche breakdown occurs.

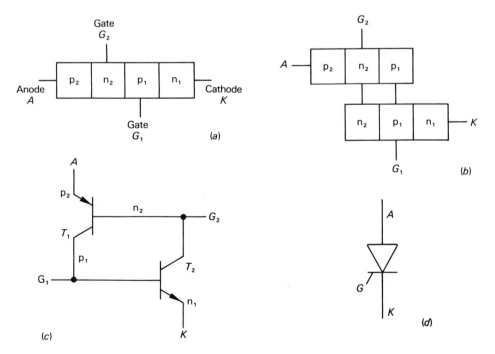

Fig. 1.25. Thyristor; (a) four-layer representation, (b) and (c) the two transistor analogy, (d) symbol.

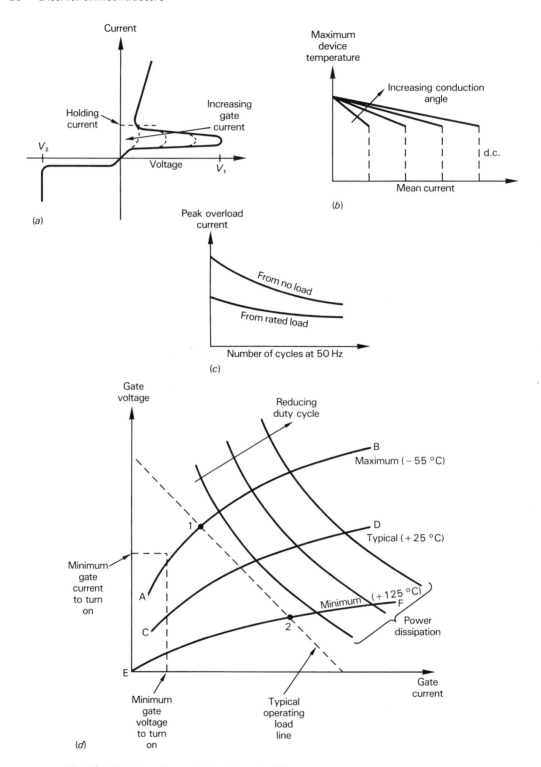

Fig. 1.26. Thyristor characteristics; (a) static, (b) mean current, (c) surge current, (d) gate.

In the forward direction the thyristor also blocks voltage until it breaks down into conduction at $V_1$. Thereafter its characteristics are similar to those of a diode provided the anode to cathode current is greater than the holding value. The larger the value of gate current the smaller the voltage at which the thyristor breaks down into forward conduction. When the thyristor is fired its gate current is made very large so that the device switches rapidly into conduction at a relatively low anode voltage.

The following are some of the ratings of a thyristor.

(i) Forward and reverse voltage rating. This is specified in two ways in data sheets. The maximum continuous rating defines a limit which, if exceeded would cause the thyristor to break down but not result in any structural damage. The peak or surge voltage on the other hand, should not be exceeded since not only would it cause the device to break down but permanent damage would also result.

(ii) Rate of change of voltage or $dv/dt$. This causes current flow through the stray anode-gate capacitances.

(iii) Current ratings. The thyristor has a mean half cycle and a d.c. current rating similar to that of a diode. For a thyristor, the conduction angle can be less than a half cycle so that the device will have a series of current ratings. These are shown in fig. 1.26$b$ where the effect of device temperature is also indicated. For short periods the thyristor is capable of a much higher current and this is shown by its surge rating of fig. 1.26$c$. If the surge is applied from no load the thyristor is cooler and can consequently carry more current.

(iv) Rate of rise of current or $di/dt$. Gate current causes a gradual spread of the turned on area of the silicon chip. Therefore, if the anode current is allowed to build up too rapidly it will result in current crowding through a small area of the device causing localised heating and burn out.

(v) Gate ratings. These determine the maximum reverse voltage which can be applied between gate and cathode and the peak gate current.

Thyristor characteristics include parameters such as the maximum volt drop at a given current, the maximum holding current, the maximum reverse leakage current, and the thyristor turn on and turn off times. These times should be as short as possible since during both these periods the current and voltage through the device are simultaneously large resulting in considerable power dissipation. The turn on and turn off times determine the maximum thyristor operating frequency.

The gate characteristics are important in the design of the thyristor drive circuitry. These are given by means of the curves shown in fig. 1.26$d$. The gate-cathode diode curves have a spread between AB and EF so that for any load line, as shown, the operating point can be situated between 1 and 2. These must be outside the box bounded by the minimum turn on voltage and current while at the same time being less than the relevant power dissipation curve. The shorter the gate duty cycle the larger the permitted gate drive, and high power pulse firing is often used for thyristors to ensure rapid turn on.

### 1.5.2 Thyristor configurations

Several different thyristor structures are used to achieve various performance parameters. Fig. 1.27$a$ shows a conventional arrangement. The edges of the silicon chip are bevelled to reduce stress at the junctions and so enable the voltage rating to be increased. An alternative technique for increasing voltage rating is to increase the thickness of the control layer, but this also results in an increased voltage drop across the thyristor. Fig. 1.27$b$ shows what is known as the *shorted emitter* thyristor structure. It is used for applications which require a high $dv/dt$ rating. In this structure the current generated by a rapidly rising voltage flows directly to the cathode so that only a small proportion of it crosses the pn junction as gate current.

It was mentioned earlier that when a thyristor turns on the initial conducting area is localised around the gate lead and then spreads relatively slowly over the whole silicon chip. The velocity of spread is about 50 to 100 m/s and since a modern, 2000 A mean rated, thyristor would have a chip diameter approaching 100 mm the turn on time of the whole chip is relatively long resulting in current crowding and a limit on the $di/dt$ rating of the thyristor. Where a large value of this rating is required special constructional techniques must be used. The most direct is to

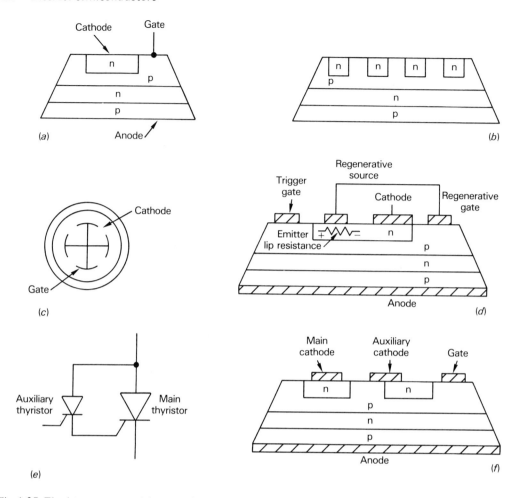

Fig. 1.27. Thyristor structures; (a) conventional, (b) shorted emitter, (c) interdigitated gate, (d) regenerative gate, (e) amplifying gate schematic, (f) amplifying gate structure.

use an interdigitated gate structure, as shown in fig. 1.27c so that several areas around the chip periphery are triggered simultaneously. Unfortunately this also means that a much larger gate current is required since the gates are in effect connected in parallel. This disadvantage is overcome in the *regenerative gate* and *amplifying gate* structures. In the regenerative gate arrangement (fig. 1.27d) conduction is commenced by a signal on the trigger gate, which is the only gate terminal brought out of the package. Once anode to cathode current starts to flow it causes a voltage loss in the cathode layer across what is known as the emitter lip resistance. This voltage drop is picked up by the regenerative source and fed to the regenerative gate which is usually an interdigitated arrangement. The external circuit needs to provide only a modest gate signal at the trigger terminal since the regenerative gate signals are derived from the load current. The amplifying gate works on a similar principle as shown in figs. 1.27e and f. Only the auxiliary thyristor is triggered externally and this provides the heavy gate drive for the main thyristor.

### 1.6 The triac

A triac has a five-layer structure as shown in fig. 1.28. It operates like two antiparallel thyristors and can be fired in quadrants I and III (fig. 1.28c).

It differs, however, from a thyristor since it can be triggered by current flowing into (plus) or out of (minus) the gate. Therefore the triac can operate in modes I (plus), I (minus), III (plus) and III (minus). For mode I (plus) terminal $L_2$ is positive with respect to $L_1$ and layers $p_1 n_2 p_2 n_3$ give the conventional thyristor arrangement with the gate at $p_2$. For mode I (minus) junction $p_2 n_3$ between $L_1$ and $G$ is forward biased. This results in electron emission into $n_2$ and this acts as a trigger current to turn on the $p_1 n_2 p_2 n_3$ layers. In mode III terminal $L_1$ is positive and layers $p_2 n_2 p_1 n_1$ are operative. For mode III (minus) the $p_2$ layer at $L_1$ and the $n_3$ layer at the gate are forward biased and give the electron emission required to turn on the device. For mode III (plus) the $p_2$ layer at the gate and $n_3$ layer at $L_1$ provide this electron emission.

The ratings and characteristics of a triac are very similar to those of a thyristor.

## 1.7 The gate turn off switch

Another controllable power semiconductor which is commonly used is the gate turn off switch (GTO). It is similar to a thyristor which is turned on by a positive gate signal but is turned off by a negative gate drive. In the simple two transistor analogy of fig. 1.25c, if the $G_1$ voltage is taken low enough it will divert all the base current from $T_2$ and so cause the pair to

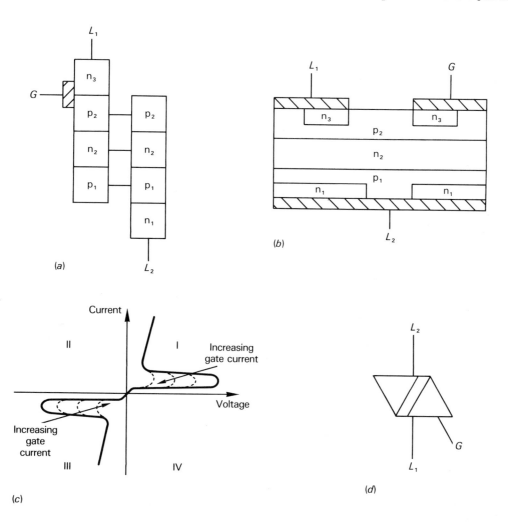

Fig. 1.28. Triac; (a) five-layer representation, (b) construction, (c) characteristic, (d) symbol.

switch off. The device is now operating as a GTO which needs a positive pulse for turn on and a negative pulse for turn off.

If the voltage across a conducting thyristor is reversed, in order to turn it off, the outermost junctions recover due to a flow of reverse current. The central junction, however, recovers relatively slowly due to the process of recombination. This is often speeded up in fast recovery devices by gold doping which reduces the carrier lifetimes. (A GTO has a faster turn off time than the thyristor since it can remove charge directly from the central junction by means of the gate contact) However it is important that excess charge is removed simultaneously from the entire junction area. If any part of the junction is not depleted of excess carriers then an attempt to turn off the GTO will cause all the load current to concentrate into this area so that it overheats. Simultaneous charge removal is accomplished by special gate designs.

An important parameter in a GTO is its turn off gain. This is the ratio of the anode current in the GTO to the negative gate current required to turn it off. Designing a GTO for high turn off gain usually increases the voltage drop across the device or reduces the breakdown voltage and makes the GTO sensitive to spurious gate signals. Due to these conflicting requirements a GTO is generally designed with a higher voltage drop and latching current than a thyristor and also with poorer gate drive and $dv/dt$ characteristics. However, it has two advantages, firstly it has a faster turn off time and secondly it can be turned off by means of a gate signal.

(Packages for thyristors, triacs and gate turn off switches are very similar to those for diodes. Low power devices are available in plastic or TO-5 cases. For larger power dissipations stud packages are used. For very large sizes press pak assemblies are common since they allow cooling for both sides of the semiconductor chip. The leads can no longer be soldered onto the die since the solder would melt due to the generated heat. Brazing or pressure contacts are used which provide better electrical and thermal contacts.

## 1.8 Trigger devices

This section describes semiconductors which are low power devices themselves but are commonly used to control high power components such as thyristors and triacs. Fig. 1.29 shows one such device, the unijunction transistor (UJT). In its simplest arrangement, the series bar structure, it consists of an n type silicon bar with two ohmic contacts for the base 1 and base 2 terminals. The emitter terminal is a p type aluminium wire which is alloyed onto the silicon bar to form a pn junction. The symbol for the UJT shows base 1 and base 2 terminals at right angles to indicate that these are non-rectifying ohmic contacts. The emitter is shown by an arrow since it is a pn junction and it points from the p type emitter to the n type base.

With zero bias on the emitter the impedance between base 1 and base 2 is high, being that of the lightly doped n type silicon bar. As the emitter voltage is raised it injects holes into the silicon bulk and these start to draw electrons from the $B_2$ terminal lowering the resistance between the emitter and $B_2$. This is also accompanied by a reduction in the emitter diode voltage on the base side. The emitter therefore becomes forward biased to a greater extent and injects more holes, resulting in a regenerative or negative resistance effect. This is shown clearly in the UJT characteristics shown in fig. 1.29e. In the cut-off region the emitter diode is reverse biased. The emitter current increases until it reaches the peak point when regeneration begins. After the valley point the incremental resistance again becomes positive.

Two other structures are shown for the UJT in fig. 1.29. The cube arrangement (fig. 1.29b) reduces the distance between emitter and base 1 and therefore gives a smaller active area. The planar assembly allows lengths to be accurately controlled giving shorter distances between emitter and base 1 and smaller chip sizes.

All junctions are also protected by a silicon dioxide layer, which results in the UJT having a faster and more uniform turn on time and a lower emitter reverse leakage current.

Unijunction transistor data sheets often give the interbase and the emitter to base 1 and base 2 resistances. The emitter to base 1 resistance varies with the value of the emitter current. The emitter to base 2 resistance is fairly constant. The latter is responsible for most of the heating when the UJT is in saturation but this can be reduced by using an external resistor in the $B_2$ lead. Fig. 1.29f shows a typical UJT trigger circuit. It operates as a free running oscillator with frequency determined by the timing cir-

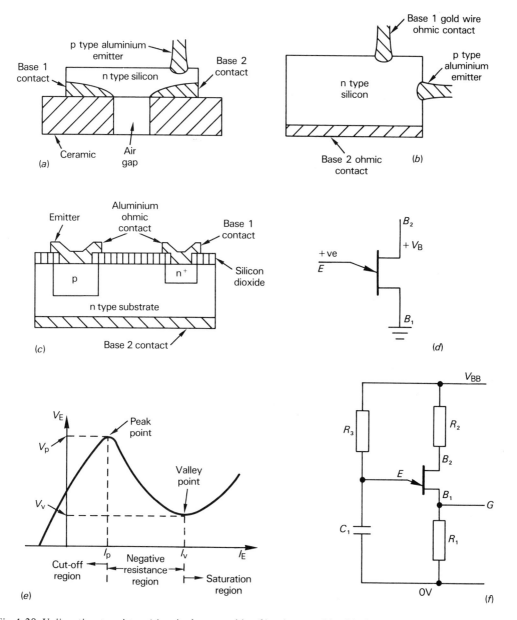

Fig. 1.29. Unijunction transistor; (a) series bar assembly, (b) cube assembly, (c) planar diffused assembly, (d) symbol and biasing, (e) characteristic, (f) oscillator circuit.

cuit $R_3C_1$. Capacitor $C_1$ discharges into $R_1$ when the UJT enters the negative resistance mode and this provides trigger pulses at G.

Peak and valley point currents and voltages are given in the data sheets and these define the negative resistance region. The intrinsic stand-off ratio $\eta$ is also given. It is defined by

$$V_P = \eta V_B + V_D \quad (1.2)$$

where $V_B$ is the $B_2$ to $B_1$ supply voltage and $V_D$ is the forward voltage drop of a silicon diode. The value of $\eta$ determines the UJT firing point for a given supply voltage. It is relatively constant with changes in $V_B$ and temperature.

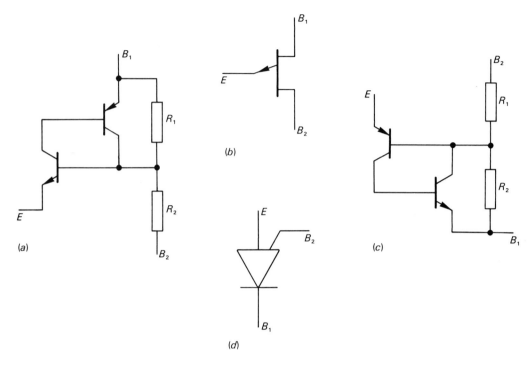

Fig. 1.30. Complementary and programmable unijunction transistors; (a) Complementary unijunction transistor (CUJT) construction, (b) CUJT symbol, (c) Programmable unijunction transistor (PUT) construction, (d) PUT symbol.

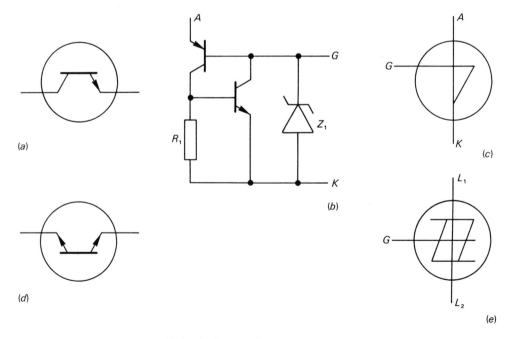

Fig. 1.31. Trigger devices; (a) trigger diode, (b) silicon unilateral switch construction, (c) silicon unilateral switch symbol, (d) diac, (e) silicon bilateral switch symbol.

Fig. 1.30 shows variations of the conventional UJT. The complementary UJT (CUJT) is really a four-layer device consisting of a pnp–npn arrangement with internal biasing resistors. The transistor pair will turn on when the emitter goes more negative than $B_1$ by a value given by (1.2). Since $\eta$ is now determined by the ratio of $R_1$ to $R_2$, and this is situated on the silicon chip, it can be more tightly controlled than for a conventional UJT. The programmable UJT is also four-layered but it has resistors $R_1$ and $R_2$ external to the chip. They can therefore be selected to programme any value of $\eta$. As seen from its symbol the PUT is basically a thyristor with an anode gate.

Fig. 1.31 illustrates further types of trigger devices. The trigger diode can have a three-layer gateless transistor structure or a four-layer thyristor arrangement. The silicon unilateral switch (SUS) is primarily an integrated circuit in which the trigger point of the two transistors is determined by the zener diode, denoted by $Z_1$ in fig. 1.31$b$. Trigger diodes which break down in both directions can be made from the double emitter arrangement of fig. 1.31$d$ or from a five-layer gateless triac structure. This is called a diac. A silicon bilateral switch (SBS) is also available. It basically consists of two SUS devices connected in reverse arrangement within the same case.

# 2. Optoelectronic components

## 2.1 Introduction

Most of the components described in this chapter are semiconductors. However, their prime function is to detect, emit or reflect optical radiation so they are classed as optoelectronic devices. The terminology and units used in describing and measuring their parameters are often confusing to electronic engineers. The terminology is briefly introduced in section 2.2. The rest of the chapter gives detailed descriptions of practical devices.

## 2.2 Optical terminology

Light is electromagnetic radiation which is a form of radiant energy. The velocity ($c$) of light through space is approximately $3.00 \times 10^8$ m/s and it is related to the frequency ($f$) and wavelength ($\lambda$) by the relation

$$c = f\lambda \qquad (2.1)$$

The velocity of light through space is constant for all types of electromagnetic radiation. Different types of electromagnetic radiation have different frequencies and thus wavelengths by (2.1), and all the different types make up what is known as the electromagnetic spectrum.

If a beam of light has a single frequency, or in practice a very narrow band of frequencies, it is said to be *monochromatic*. If the individual waves making up this beam all have the same phase relationship they are said to be *coherent*. The waves move back and forth at right angles to the direction of travel of the beam of light. Ordinarily these waves are not pointed in any given direction in this plane and the light is said to be unpolarised. It is possible, by techniques to be described later, to change the light such that the wave motion is in one direction only in a plane perpendicular to the direction of motion. It is then called linearly *polarised light*. Other forms of polarisation such as circular or elliptical also exist but are not as common as linear.

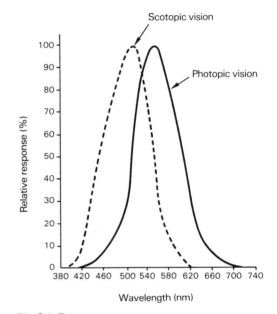

Fig. 2.1. Eye response curves.

There are two units in common use for measuring a quantity of light, the radiometric and the photometric. The *radiometric unit* gives a measure of the radiation of all wavelenths within the optical spectrum, and is therefore concerned with total radiation detection. The *photometric unit,* on the other hand, is primarily concerned with the measurement of visible light, i.e. between the wavelengths of $0.38 \times 10^{-4}$ cm and $0.76 \times 10^{-4}$ cm, and with the response of the human eye to this radiation. However, the eye has two types of response as shown in fig. 2.1. At normal illumination levels the response is *photopic*, but at low light levels it is *scotopic* (which is insensitive to colour). Usually the photopic curve is used for photometric measurements, and this peaks at a wavelength of $0.555 \times 10^{-4}$ cm.

Fig. 2.2 illustrates two geometrical terms which are used to define optical units. The

## 2.2 Optical terminology

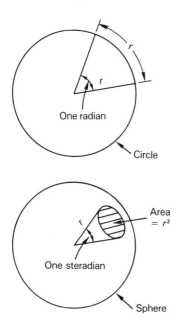

Fig. 2.2. Illustration of a radian and a steradian.

*steradian* is the solid angle formed at the centre of the sphere of radius $r$ by the area of $r^2$ on the surface.

The terminology used for radiometry and photometry are summarised in fig. 2.3 and are similar except that the word 'radiant' is used for radiometry and 'luminous' for photometry.

In radiometry the energy of the travelling optical wave is measured in joules and is called radiant energy. If this energy is measured in a unit of time it is called radiant power and is given in units of joules per second or watts. The energy emitted from a point source of light, called the radiant intensity, can be conveniently measured in watts per steradian. Two terms can be used to define the light energy incident on a surface. The first is radiance, measured in watts per steradian square metre. The second is irradiance, measured in watts per square metre. Light energy reflected or transmitted through this surface is called radiant exitance and is measured in watts per square metre.

In photometry the wave energy is called luminous energy and is measured in lumen seconds. The energy per unit time is the luminous power, measured in lumens. The energy from a point source is luminous intensity given as lumens per steradian which are also called candelas. The light energy falling on a surface may be measured in lumens per steradian square metre, equivalent to candelas per square metre, or in lumens per square metre. The first

| Parameter | Radiometric | | Photometric | |
| --- | --- | --- | --- | --- |
| | Name | Unit (and abbreviation) | Name | Unit (and abbreviation) |
| Wave energy | Radiant energy | joule (J) | Luminous energy | lumen second (lm/s) |
| Wave energy per unit time | Radiant power | joule/second (J/s) or watt (W) | Luminous power | lumen (lm) |
| Emitted energy from a point source | Radiant intensity | watt/steradian (W/sr) | Luminous intensity | lumen/steradian (lm/sr) or candela (cd) |
| Energy on a surface | Radiance | watt/steradian metre$^2$ (W/sr m$^2$) | Luminance | lumen/steradian metre$^2$ (lm/sr m$^2$) or candela/metre$^2$ (cd/m$^2$) |
| | Irradiance | watt/metre$^2$ (W/m$^2$) | Illumination | lumen/metre$^2$ (lm/m$^2$) |
| Energy leaving surface | Radiant exitance | watt/metre$^2$ (W/m$^2$) | Luminous exitance | candela/metre$^2$ (cd/m$^2$) |

Fig. 2.3. Summary of radiometric and photometric terms and units.

unit is called luminance and the second is illumination. The energy leaving the surface is the luminous exitance and is measured in candelas per square metre.

## 2.3 Optical sources

Light sources range from the conventional incandescent and discharge lamps to the newer *light emitting diodes*. The different types of lamps are briefly introduced in this section and the light emitting diode (LED) is then described in greater detail. Only discrete emitters, used as panel indicators and lasers, are covered, display devices being discussed in section 2.6.

### 2.3.1 *Incandescent source*

The incandescent lamp used as an indicator in electronic equipment is similar in operating principle to its household counterpart but is physically much smaller, having bulb sizes down to 3 mm in diameter. A tungsten filament is enclosed in a glass envelope and gives off light when a current is passed through it to raise it to incandescence. The filament can be designed to run from a wide range of supply voltages varying from 1 V to 60 V a.c. or d.c. Its current requirements are between 10 mA and 400 mA. The light output is in the visible spectrum and since it is white light it can be filtered to produce any colour. The maximum radiant intensity is in the range 0.05 to 300cd and the lamp can be dimmed by varying the filament current. The incandescent lamp can produce a high intensity output which may be used for flood lighting. It has a life of about 25 000 hours, but has the disadvantage of being unable to withstand shock and vibration or to operate in a pulsed mode. It is also physically larger than other lamps. The efficiency of the lamp, in common with other light sources, is low. Only about 2% of the electrical input is converted to visible radiation and if filters are used this can fall as low as 0.2%.

### 2.3.2 *Halogen cycle source*

The halogen cycle lamp is similar to an incandescent lamp but is made from an envelope of material such as quartz, which is capable of withstanding high temperatures, and contains halogen vapour. This vapour combines with particles of evaporated tungsten and redeposits them onto the filament so that bulb blackening is avoided. The lamp can be made in small sizes, down to about 10 mm bulb diameter, and can be designed to operate over a wide range of voltages and currents in the range 1 V to 60 V and 10 mA to 10 A. The light output is similar to that of an incandescent lamp although the life is shorter. The halogen cycle lamp is also not very good for operating over extremes of temperature.

### 2.3.3 *Gas discharge source*

Another light source used in electronic equipment is the glow or discharge lamp. It consists of two electrodes sealed in an envelope containing a gas such as neon. At a sufficiently high voltage, current flows through the gas ionising it and causing a glow near the negative electrode. The glow lamp requires a high operating voltage, greater than 100 V, but this can be a.c. or d.c. and the current input is low, less than 2 mA. It has a life of about 25 000 hours and its failure mode is a deterioration of the light output rather than catastrophic failure. Glow lamps are available in a variety of colours, operate over wide ranges of temperature and are good at withstanding shock and vibration.

### 2.3.4 *Light emitting diodes*

The light emitting diode (LED) is based on the phenomenon of *electroluminescence,* which is the emission of light from a semiconductor under the influence of an electrical field. The phenomenon was first observed in 1907 but it was not until 1964 that the first commercial device was produced. Electrons in a solid exist in one of two states, either bound to the nucleus or free in the lattice. The energy in the free state is greater than in the bound state. The conduction band in a semiconductor is the energy band for free electrons. The valence band is the energy band for bound electrons or positive holes. The energy gap or band gap or forbidden gap is the energy difference between these two bands. Transition of electrons between the two bands may be direct or, if the electron moves via a process of recombination, indirect. Indirect transitions have a lower *quantum efficiency*.

An LED needs a plentiful supply of electrons in the conduction band and these can be provided by raising the energy of the valence band electrons. These electrons then give up their energy in the form of photons of light and return to the valence band. The excitation process can be conveniently obtained by forward biasing the pn junction of the diode.

The wavelength of light emitted by an LED is directly related to its energy gap, $E$. If $h$ is Planck's constant, $c$ the velocity of light and $\lambda$ the wavelength of the emitted radiation then

$$\lambda = hc/E \qquad (2.2)$$

On the basis of (2.2) materials which have a band gap energy between 1.65 eV and 3.2 eV will be capable of emitting visible light. However, since electrons often move in stages within the forbidden gap they also give up their energy in steps, so that the emitted wavelength may be longer than that given on the basis of the band gap.

In the preparation of semiconductors such as gallium phosphide (GaP) for LEDs, Czochralski growth techniques are used to pull the material from its melt. Liquid encapsulation is used, and since the partial pressure of phosphorus will be high on its free surface, the loss of vapour pressure components from the melt is suppressed by means of a flux layer, such as boric oxide ($B_2O_3$), on top of the melt, and an inert gas such as argon or nitrogen at high pressure. For good quality devices the pulled crystals are used as the substrates and a layer of epitaxy is grown on them. Two techniques exist for epitaxy preparation, namely the *vapour phase* and *liquid phase* techniques.

In vapour phase epitaxy (VPE) either the constituent elements are transported over the substrate in the form of compounds and deposit a layer on it by chemical reaction, or the material itself is carried in a stream of gas and deposited on the surface. In both cases the epitaxy can be doped by introducing the impurity into the vapour stream.

In liquid phase epitaxy (LPE) the constituents of the required compound are dissolved in a suitable material. The substrate can now be dipped into the melt and then cooled to form the epitaxy, or alternatively the melt can be passed over the stationary substrate. A series of different melts can be used to build up a complex epitaxy structure and each melt can be doped. Liquid phase epitaxy is a lower temperature process than vapour phase so that it produces a lower crystal dislocation and defect concentration. However, the vapour phase process is more versatile since parameters such as layer thickness and dopant concentration can be varied easily. The pn junctions in the epitaxy can be either grown or formed by diffusion techniques. For liquid phase epitaxy they are usually grown and for vapour phase epitaxy they are usually formed by a zinc diffusion.

Several different materials are used in the preparation of LEDs, only a few of these being described here. Many considerations govern the choice of LED material. The energy gap must be suitable for the wavelength required. The efficiency of emission is also important. Most of the materials presently used are based on compounds which are prepared from equal atomic proportions of group III and group V elements. For infrared emitters gallium arsenide is the most popular material. It is a group III–V compound having a direct energy gap of 1.43 eV, and a wavelength of emission of around 900 nm.

Gallium phosphide is an indirect gap III–V

| Material | Colour | Peak wavelength (nm) | External quantum efficiency (%) | Band structure | n layer growth | p layer growth | Lumens per watt |
|---|---|---|---|---|---|---|---|
| GaP : Zn,O | Red | 695 | 15.0 | Indirect | LPE | LPE | 20 |
| GaP : N | Green | 570 | 0.5 | Indirect | LPE | LPE | 600 |
| GaP : NN | Yellow | 590 | 0.1 | Indirect | VPE | Zn diffusion | 450 |
| $GaAs_{0.6}P_{0.4}$ | Red | 649 | 0.5 | Direct | VPE | Zn diffusion | 75 |
| $GaAs_{0.35}P_{0.65}$ : N | Orange | 632 | 0.5 | Indirect | VPE | Zn diffusion | 190 |
| $GaAs_{0.15}P_{0.85}$ : N | Yellow | 589 | 0.2 | Indirect | VPE | Zn diffusion | 450 |

Fig. 2.4. Summary of the properties of light emitting diode (LED) materials.

compound having a gap energy of 2.26 eV. It is used for emission in the visible spectrum. It can be doped with zinc and oxygen to give red light or with nitrogen to give green. As the concentration of nitrogen increases the emission moves from a peak at 560 nm (green) to 585 nm (yellow). Gallium arsenide and gallium phosphide combine to form a solid solution of gallium arsenide phosphide. The light emission from this material peaks at about 660 nm (red). Nitrogen doping increases the conversion efficiency of the phosphorus and also the wavelength of the emission.

Fig. 2.4 summaries the properties of some of the LED materials for the visible spectrum. The external quantum efficiency is a measure of the power output. Luminous efficiency in lumens per watt considers the sensitivity of the eye. The eye is more sensitive to green than to red hence the increase in luminous efficiency in this region.

### 2.3.5 LED characteristics

Electroluminescent diodes emit radiation in relatively narrow spectral bands so they can be designed to give a range of colours without filters. The intensity of the emitted light decays exponentially with on time so that it is convenient to quote the life of the diode as its *'half life'* which is the time for the luminance level to fall to half its original value. The effect of temperature on an electroluminescent diode relates primarily to variations in its band gap. As the temperature increases the wavelength at which the light peaks also increases. This causes a shift in the colour and since this affects the eye response the luminous efficiency also changes. For red and green diodes the efficiency decreases with temperature, but for yellow diodes it increases since the emitted wavelength moves more into line with the eye response.

Electrically, the LED has very similar characteristics to a conventional diode. In the forward direction it exhibits a 'knee' in the voltage-current curve between about 1.1 V to 4 V. The knee occurs at a higher voltage as the wavelength decreases since the energy gap increases. In the reverse direction the maximum voltage is usually specified at a low value in order to keep the leakage current small and so prevent degradation of the lamp characteristics.

The LED consumes comparitively little power, is small in size, rugged, and can withstand shock and vibration. It has a very fast

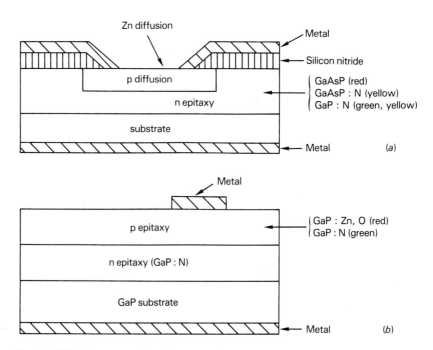

Fig. 2.5. LED chip construction; (a) vapour phase epitaxy, (b) liquid phase epitaxy.

response time, of about 10 ns, and can be dimmed by adjusting its mean current. It is available in red, green, yellow, orange and infrared. It can only operate from d.c. and is not able to withstand extremes of temperature.

### 2.3.6 Structure of the LED

The structures of VPE and LPE LED chips are shown in fig. 2.5, which indicates the dopants required to produce the different colours. If the refraction index of a semiconductor chip is $n_s$ and of air $n_a$ then the reflection coefficient $r$ at the air-semiconductor interface is given by

$$r = \left(\frac{n_s - n_a}{n_s + n_a}\right)^2 \quad (2.3)$$

and the critical angle $\theta$ is given by

$$\sin \theta = n_a/n_s \quad (2.4)$$

For an LED chip, $n_s$ is relatively large so that the reflection coefficient is high and the critical angle is small. This means that a very small solid angle exists within the chip from which external radiation can be obtained. If a transparent material having a refraction index greater than $n_a$ and less than $n_s$ is placed over the LED, then more radiation will go from the chip to the material. Ideally the coating should have a refractive index of $(n_a n_s)^{1/2}$. The covering material can be designed as a hemispherical dome so that light entering it from the chip will strike the hemispherical surface at near normal incidence and therefore pass through without internal reflection.

Fig. 2.6 shows the construction of a typical LED lamp. Light is emitted in all directions from the pn junction, and internal reflections also cause a large proportion of it to reach the back surface of the chip. The copper lead is therefore designed to act as a reflector of this light as well as heatsinking the chip. The epoxy encapsulation can be coloured to act as a filter and let through emitted light while cutting down reflected background light, so increasing the contrast between emitted and reflected light. The chip covering can also be formed into a lens so making the lamp into a directional indicator. The relative intensity against angular displacement plot for an encapsulated LED is called its *polar diagram*, a typical plot being shown in fig. 2.7.

### 2.3.7 The LASER

A light source which is finding increasing commercial applications is the LASER, which stands for Light Amplification by Stimulated Emission of Radiation. The essential properties of a laser beam is that it has high radiant intensity, it can be polarised, and it is monochromatic and coherent. The laser can generally be looked on as an optical oscillator, that is an amplifier with positive feedback. Its three main parts are

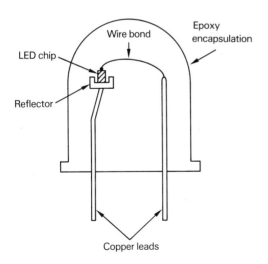

Fig. 2.6. LED lamp construction.

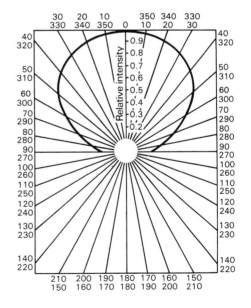

Fig. 2.7. Polar diagram for a typical encapsulated LED.

an active material which gives gain, a method of exciting the active material, and a resonant structure to produce feedback. Many materials are used in lasers and these may be solid, liquid or gas.

A semiconductor laser consists of a forward biased pn junction which is excited by an electric current. Near the junction recombination of holes and electrons takes place releasing photons. If the bias current is large enough then many holes and electrons are concentrated in a small area of the junction and the photons which are released will stimulate more emission. Relatively high current densities are required to produce lasing and many laser structures have been developed to reduce this current requirement and to increase life. A few of these are illustrated in fig. 2.8. The homojunction laser is the oldest and simplest. It can be made with a p diffusion or an epitaxy. Heterojunction lasers all aim to confine the optical radiation to the junction region. The narrower the junction the lower the current density needed for lasing, but narrow junctions also make the laser more susceptible to catastrophic failure. The double heterojunction confines radiation to a narrower region than the single heterojunction so that lower current densities are required, but the device is also more susceptible to failure. The LOC structure controls the width of the lasing junction by using a thin p type recombination layer and wider n type layer for light propagation. This aims to reduce current densities without adversely affecting life.

## 2.4 Optical detectors

Photodetectors can be junctionless or 'bulk'' types, such as the *light dependent resistor* (LDR), or they can have junctions such as photodiodes, phototransistors and photothyristors. The detectors can also be discrete or several can be integrated onto a silicon die, usually with other electronic circuitry.

### 2.4.1 *Light dependent resistors*

Photoconductive or photoresistive devices, known as light dependent resistors (LDRs), are made from materials whose resistance decreases with exposure to light i.e. whose conductivity increases with the illumination level. The light energy causes hole-electron pairs to be generated and the electrons jump into the conduction band. All semiconductors exhibit this phenomenon but within the devices several processes are in

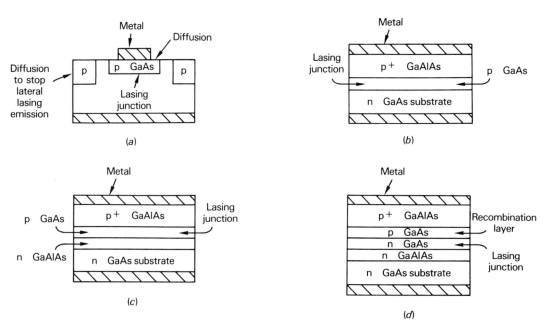

Fig. 2.8. Semiconductor laser structures; (*a*) homojunction, (*b*) single heterojunction, (*c*) double heterojunction, (*d*) large optical cavity (LOC).

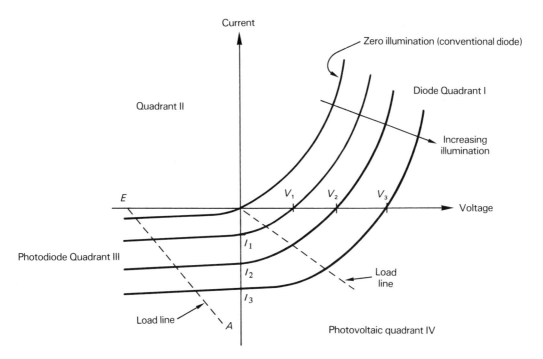

Fig. 2.9. pn junction photodetectors.

competition, such as recombination and trapping centres, and the generation of thermal hole-electron pairs. In an LDR the photoconductive effect is enhanced by a special doping of the semiconductor. The materials most commonly used are an aluminium oxide ceramic substrate which is coated with a layer of cadmium sulphide (CdS) or cadmium selenide (CdSe). The devices are made by a variety of techniques such as sintering, firing or by vapour, chemical or vacuum deposition.

The spectral response of an LDR extends over the visible and infrared region. The response of the cadmium sulphide device peaks at 550 nm while that of the cadmium selenide device peaks at 700 nm. Photoconductors have a large sensitive area, large ratio of dark to light resistance, and zero offset voltage. They cost relatively little and can be used bidirectionally. They also have the advantages of a high working voltage and large power dissipation. The disadvantages of photoconductive devices are that they are slow, with a response time of many milliseconds, and they exhibit a hysteresis effect in their resistance-illumination characteristic. This means that the conduction is a function of the cell's previous history of exposure to light intensity and duration. The hysteresis effect is most noticeable when the LDR is used at low levels of illumination. Due to their low cost and high sensitivity LDRs are used in applications such as street lighting, camera exposure control and industrial counting where operating speed is not high.

### 2.4.2 Photodiodes

Fig. 2.9 shows the characteristics of a single junction device. With no illumination it behaves like a conventional diode giving a low voltage drop in quadrant I and a low leakage current when reverse biased in quadrant III. As the illumination level increases it gives rise to a series of curves, as shown. If the photodiode is reverse biased by an external supply $E$ and connected to a load then it will operate in quadrant III, since both current and voltage are reversed, and the load line is $EA$. If no external supply is connected to the diode then the incident light will create a potential across the junction such that the type p layer acquires a positive potential relative to the n type layer. This will cause a current to be driven through an external load resistor such that the voltage is positive but the current is reversed through the diode. The device is now called a photovoltaic diode and operates in quadrant IV. In fig. 2.9 voltages $V_1$, $V_2$ and $V_3$ are the open circuit voltages of the photovoltaic diode and $I_1, I_2, I_3$ are its short-circuit currents.

The photodiode works on the principle that in the absence of light the current through the reverse biased pn junction, called the *dark current,* is very small. Incident light creates hole-electron pairs within the junction and these are carried across the junction by the external bias. The effect of the external bias is to increase the depletion region so that the device is very efficient in converting light to electrons, and gives a linear current-light characteristic.

Several parameters are important in the design of a photodiode circuit. The junction capacitance should be low since the speed of response is primarily determined by this capacitance and the load resistance. The capacitance is proportional to the junction area divided by the square root of the reverse voltage, However, the diode noise and dark current also increase with voltage so that a compromise is often required. The dark current of the diode is proportional to the device area, the temperature, and the square root of the voltage, and should be as small as possible.

The performance of a photodiode can be measured in terms of the quantum efficiency, which gives a measure of the electron emission for a quantum of incident light. Photodiodes have the highest efficiency of all solid state detectors, reaching 95%. Alternatively the efficiency of a photodiode can be measured in terms of its responsivity which is the current produced per watt of light irradiance. Responsivity is a function of many parameters such as wavelength, applied bias voltage, frequency and temperature.

The noise generated in a photodiode is measured in terms of its *noise equivalent power* (NEP). This is the amount of light needed to produce a signal equivalent to the noise level and takes into account the quantum efficiency and noise. Diode noise, however, increases with junction area so that NEP is worse with large area devices. Detectivity is defined as NEP per active area and so is a parameter which is independent of area.

Photodiodes are usually made from silicon. The diodes can be designed to optimise parameters such as speed, dark current, and spectral response by varying process factors such as material resistivity, type of dopants, and diffusion depth. However, different structures have been developed for special applications. For example, in a conventional photodiode at very low load resistances, the operating speed is limited by the diffusion time of the carriers in the heavily doped regions and their transit times in the space charge region. pin diodes, which have an intrinsic layer between the p and n regions, give a lower junction capacitance and can operate much faster. The dominant effect for a pn junction is the diffusion current whereas for a pin it is the drift current. For very high speeds avalanche photodiodes (APD) may be used. These operate at high reverse voltages and work on the principle that the electrons generated by the incident light are rapidly multiplied due to the avalanching effect through the junction. The avalanche effect also creates noise, which is a disadvantage of the APD.

For large area detectors it is difficult to get good sensitivity over the whole device since the uniformity of the diffusion depth cannot be maintained. In such cases Schottky barrier diodes are preferred in which a thin layer of gold is evaporated onto the silicon to form the junction. Using this technique photodiodes as long as 30 cm have been built. The disadvantage of a Schottky device is that it cannot be used at high temperatures or high light levels.

Photodiodes are generally used in applications where amplifiers with high input impedances are needed. Due to their linear characteristics they are often assembled with these amplifiers into a single package, and used in card and tape readers, as isolators, and for general counting applications. The salient parameters of photodiode and photovoltaic detectors are included in fig. 2.10.

### 2.4.3 *Phototransistors*

The phototransistor is probably the most widely used optical sensor. It can be regarded as a photodiode connected to an amplifying transistor. The photodiode is formed as part of the reverse biased collector-base junction of the transistor, and this junction is designed to have

| Parameter | Photo-conductive | Photo-emissive | Silicon photovoltaic | Photodiode | Photo-transistor | Photo-thyristor |
|---|---|---|---|---|---|---|
| Maximum temperature (°C) | 75 | 80 | 150 | 125 | 125 | 100 |
| Maximum voltage (V) | 1000 | 2800 | 0.5 | 200 to 2000 | 60 | 200 |
| Type | Symmetrical | Asymmetrical | Asymmetrical | Asymmetrical | Asymmetrical | Asymmetrical |
| Maximum current | 1 A | 10 mA | 1 A | 5 mA | 50 mA | 1.5 A |
| Power dissipation | 20 W | 0.01 to 1W | 400 mW | 50 mW | 400 mW | 2 W |
| Switching times | 1 to 100 ms | 0.1 $\mu$s | 1 to 100 $\mu$s | 1 ns to 1 $\mu$s | 2 to 100 $\mu$s | 2 $\mu$s |
| Maximum frequency | 1 kHz | 10 MHz | 50 kHz | 10 MHz | 100 kHz | 1 kHz |
| Operating light level (mW/cm$^2$) | 0.001 to 70 | $10^{-9}$ to 1 | 0.001 to 1000 | 0.001 to 200 | 0.001 to 20 | 2 to 200 |
| Long term stability | Fair | Good | Best | Good | Good | Good |
| Peak spectral response ($\mu$m) | 0.6 | | 0.85 | 0.85 | 0.8 | 0.85 |
| Size | Medium | Large | Medium | Smallest | Small | Small |

Fig. 2.10. Comparison of photodetector properties.

a large area to enhance its efficiency as a photodetector.

The phototransistor behaves very much like a conventional transistor except that its base current is controlled by optical radiation. The collector current is approximately equal to the collector-base photocurrent multiplied by the transistor gain. The transistor gain is different depending on the process of manufacture used and varies with the magnitude of the collector current so that the phototransistor has a relatively non-linear collector current-illumination characteristic.

The phototransistor has between 100 and 1000 times more sensitivity than a photodiode due to its internal amplification. However, the dark current is amplified as much as the photocurrent so that signal to noise ratio is no better. The phototransistor is also slower than a photodiode partly due to the *parasitic collector-base capacitances,* and partly due to limitations in current gain–bandwidth product. For high speed operation, smaller area devices may be used but then the sensitivity is also correspondingly lower.

For higher sensitivities, in excess of about 10 000 times that obtained from a photodiode, a photo Darlington device may be used. In essence this device consists of a phototransistor, which forms the input stage, and a conventional transistor, forming the output stage, both built into a single silicon die. Field effect transistor photosensors have about ten times higher sensitivity than bipolar phototransistors since the higher input impedance gives larger voltage swings from a lower gate photocurrent. The FET phototransistor uses the gate to channel junction as a photodiode. It has a very low offset voltage but is slower than a bipolar phototransistor.

### 2.4.4 Photothyristors

The construction of a photothyristor, also called a light activated SCR or LASCR, is shown in fig. 2.11. In the absence of light, junction $J_2$ is reverse biased. When illuminated, hole-electron pairs are created in the vicinity of $J_2$ and are swept across to the anode and cathode. This acts as a triggering current and if it is large

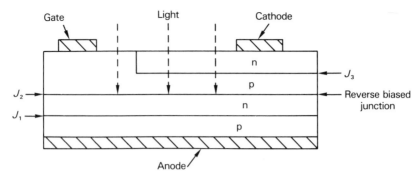

Fig. 2.11. Photothyristor construction.

enough it will turn on the thyristor. Usually the gate lead is brought out of the package and a bias can be applied to it to vary the threshold light level for turn on.

The photothyristor is generally made from thin layers to enable greater light penetration. Unfortunately this also results in a lower blocking voltage. It is also designed with a larger junction to increase light sensitivity, but this also makes it more sensitive to variations in temperature and voltage, and gives it a longer turn off time than a conventional thyristor. Resistors connected between gate and cathode of the thyristor reduce its susceptibility to noise and $dv/dt$ effects, but also reduce its light sensitivity.

### 2.4.5 *Photomultipliers*

Before leaving discrete photodevices it would be useful to consider photo-emissive devices such as the photomultiplier tube in which the anode current is proportional to the intensity of the incident light. The disadvantage of photo-emissive sensors is that they are large and expensive, and need between 300 and 2500 V for operation. Their advantage is that they have a higher frequency response than photodiodes and are more sensitive. The spectral response of photo-emissive sensors can be varied by changing the cathode material so that they can range from 100 nm to 1000 nm. Their key parameters are included in fig. 2.10.

### 2.4.6 *Integrated sensors*

Discrete sensors can be integrated with amplifying or switching circuitry onto the same silicon die. Several photosensors can also be integrated together in order to form an array. The main problem in these components is providing enough pins on the package to give access to each sensor. For large arrays some form of *multiplexing* is generally used. Fig. 2.12a shows a linear array of photodiodes which are controlled by series transistors. The signal which switches on the relevant transistor is fed serially into a shift register. The data output at any time corresponds to the diode which is being accessed. The source of the control transistor is made into the anode of the phototransistor, as shown in fig. 2.12b, giving a high level of integration.

Considerations such as spectral response, responsivity and dark current are common to discrete photodiodes and those used in arrays. In addition array photodiodes must have uniformity of response across the array. This is determined by the uniformity in the optical area of each diode, freedom from silicon defects, and uniformity in the carrier lifetime. This latter parameter affects the depth from which carriers can diffuse without recombination and mainly influences the infrared response of the array. Another factor to be considered in an array of photodiodes is cross talk. This is the response which a photodiode shows to light falling on an adjacent photodiode. Cross talk is caused by the diffusion of charge carriers in the substrate. It is worse in infrared light and it can be reduced by filtering out this radiation before it strikes the photodiode.

## 2.5 Optical couplers

Optical sources and detectors can be put into a single package, and these devices are called *optical couplers*. The gap between source and detector may be totally enclosed by the package

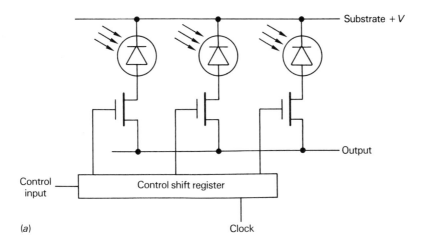

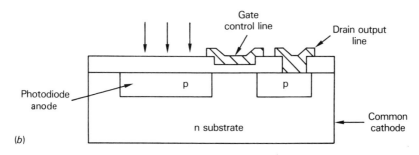

Fig. 2.12. Photodiode arrays; (a) linear array, (b) single cell of array.

or it may be accessible. Couplers in which the gap is accessible are generally used for motion sensing applications. Since source and detector are relatively wide apart the coupling efficiency between them is poor. Furthermore, the breakdown voltage between input and output may be relatively low since there is only air between them. Totally enclosed couplers use glass or plastic separators between source and detector and can therefore be placed closer together whilst still giving a relatively high isolation voltage. They are used in a wide range of applications such as solid state switching, data line driving and ground loop isolation.

The parameters of most interest in optocouplers are the isolation between source and detector, the input-output current transfer ratio and the speed of operation. The isolation resistance is of the order of $10^{11}$ $\Omega$, and is usually higher than the leakage resistance between package pins on the printed circuit board. Another way of expressing this isolation is by the maximum voltage which can be applied between input and output without breakdown. If breakdown occurs it can form a resistive path due to carbonised moulding on the surface, or it can result in a short circuit caused by molten lead wires bridging the lead frames of the emitter and detector. For high isolation voltages the moulding is usually designed such that the input and output pins are brought out from separate sides of the package. It is also important to minimise the parasitic capacitance through the dielectric between input and output.

The current transfer ratio is given as the ratio of the output current to the input current of the source, when the detector is biased in a specified way. This ratio is determined by several factors such as the level of current into the source and detector saturation. Generally, an LED is used as a source and the light output of the device falls with time giving a decrease of transfer ratio. The operating speed of the

40  *Optoelectronic components*

coupler defines how fast it can be switched, and is usually specified in terms of its maximum operating frequency.

An LED emitting in the infrared region is used as the source in most modern optocouplers while photodiodes, phototransistors, photo Darlingtons, or photothyristors are used as detectors. The photodiode gives the highest operating speed, which is typically in the range 100 kHz to 5 MHz. However the output current is very low so that the diode is usually followed by an amplifier or logic switch integrated on the same die as the photodiode, giving a current transfer ratio of between 5% and 500%. Optocouplers using phototransistors cost little and are widely used. They have typical speeds of 100 kHz to 500 kHz and a minimum current transfer ratio between 10% and 100%. Photo Darlington devices have transfer ratios between 100% and 600% but this is difficult to predict accurately due to the wide variation in the gain of the Darlington stage. The operating speed is relatively low, being typically between 2 kHz and 10 kHz.

For high currents, photothyristor output stages are used. The current into the LED, which is needed to trigger the thyristor, is now an important parameter. Since the coupling efficiency between the LED and photothyristor is low it is important that the thyristor is designed to have a high gate sensitivity. This usually requires careful process control in order not to degrade other parameters such as voltage rating. A photothyristor coupler requires between 10 mA to 30 mA input current to trigger and can provide between 200 mA to 300 mA output current. The thyristor turn on time is typically 2 µs to 20 µs.

### 2.6 Displays

There are many different types of displays and these can be grouped in various ways. In this section four groupings are used i.e. LED, liquid crystal, gas discharge and an assorted collection of devices which are not widely used.

Displays vary in area from a few square centimetres to several square metres. They can be designed to display numbers only or *alphanumeric* characters. Displays can be made of single characters or many hundreds of characters. They are available in a variety of technologies some of which compete whilst others complement one another. Since displays are intended to be viewed, their appearance is as important as their electrical characteristic, and considerable work has been done in making the display acceptable to the viewer.

Several factors need to be considered when

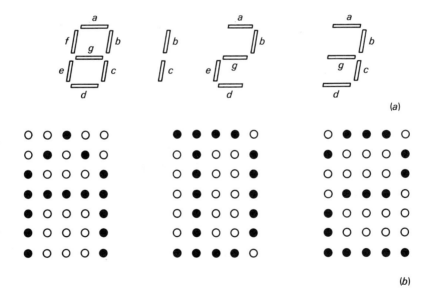

Fig. 2.13. Display formats; (*a*) 7-bar showing numerals 8, 1, 2, 3, (*b*) 7 x 5 dot matrix showing letters A, D, and numeral 2.

choosing a display. The character format is usually determined by the application. 7-segment displays, such as shown in fig. 2.13a, provide a cheap method for displaying numeric information. Extension to 16 bars gives full numeric and limited alphabetical capability. For alphanumeric displays, a *dot matrix* format is usually used. The colour of a display is also important. A surprisingly large number of displays emit in the red spectrum but many people find red tiring to view for long periods. There is much less red in ambient light than other colours such as yellow or green, and one can use cheap filters to let through the long red wavelength whilst cutting off other reflected light, so that red displays have very good contrast.

Another parameter of interest in displays is *viewing angle*. There are many ways of defining this, the most commonly used being the half power point i.e. the angle at which the luminous intensity of the display is half the value observed when the display is viewed head on. The life of a display is also important but since most displays fail by a gradual reduction in their light output it is difficult to measure. It is usually defined as the time after which the light output has been halved, all other factors remaining unchanged.

Displays can be used to indicate a variety of information. Apart from the digital displays of fig. 2.13 analogue displays can be used in which light sources are placed in a circular format, as in the dial of a watch, or are arranged in a straight line and lit in sequence to represent the increase or decrease in analogue data.

### 2.6.1 *Light emitting diode displays*

The light emitting diode display has characteristics very similar to the discrete emitters discussed in section 2.3 except that now the radiation occurs in the visible spectrum. LED displays are available in 7-bar and dot matrix constructions. They can be built by assembling discrete diodes on an insulating substrate, or by monolithic techniques in which p or n junctions are diffused into a silicon die of the opposite polarity. Monolithic techniques require less manual assembly but use more silicon, and are therefore used mainly for smaller displays. In either case it is usual to join together the

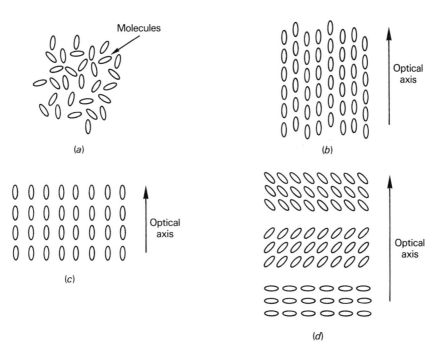

Fig. 2.14. Classes of liquid crystal states compared with an isotropic liquid; (*a*) isotropic liquid crystal, (*b*) nematic liquid crystal, (*c*) smectic liquid crystal, (*d*) cholesteric liquid crystal.

42  *Optoelectronic components*

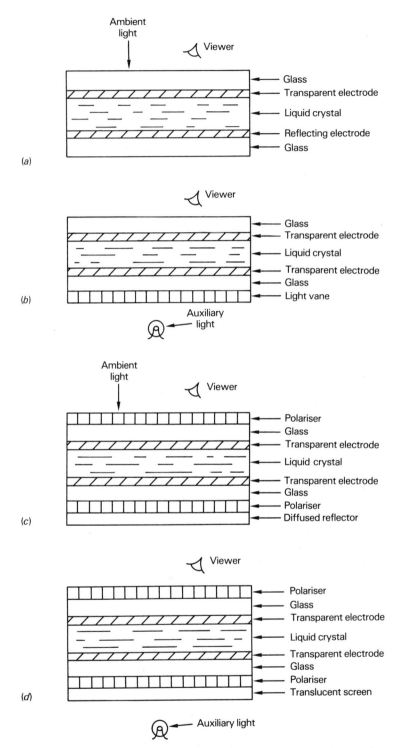

Fig. 2.15. Types of liquid crystal displays; (*a*) dynamic scattering, reflective, (*b*) dynamic scattering, transmissive (*c*) field effect, reflective, (*d*) field effect, transmissive.

cathodes or anodes of all the diodes in order to save on the number of package pins required. Small monolithic displays usually have plastic magnifiers moulded into their package. For very large displays hybrid assemblies are used in which light piping increases the effective size of each display.

The LED display is compact, resistant to shock and vibration, has a long life, is compatible with other semiconductor devices, operates with a low segment current of 0.1 mA to 5 mA and has a very fast response time of 10 ns to 1 $\mu$s. Its disadvantages are that it is susceptible to thermal failure; has a low optical conversion efficiency in the region of 1% to 5%; has limited maximum luminous intensity, which makes it unsuitable for use in high ambient lighting; and it suffers from '*washout*' in direct light, due to light reflections from the surface of the silicon die, making it difficult to read the display. The light output from the LED increases at a faster rate than the input current, until the thermal limit is reached. This means that if a LED is operated in a pulsed mode it will emit more light for the same mean current than when running from d.c.

### 2.6.2 *Liquid crystal displays*

The material used in liquid crystal displays is in an anisotropic phase between the solid crystalline and isotropic liquid phases. Many organic chemicals exhibit this property, the three parameters of most importance for displays being electrical conductivity, dielectric constant and refractive index.

Liquid crystals are generally grouped into three classes as shown in fig. 2.14. In the nematic class the rod like molecules of the liquid crystal are all parallel to each other. In the smectic material these molecules are not only all parallel but they are also stacked in parallel layers. In cholesteric liquid crystals the molecules form a spiral, the axis of the molecules being twisted about the normal to the axis in a helix. For displays the nematic and cholesteric materials are mainly used.

Liquid crystal displays are passive since they only modify the incident light and do not emit any light themselves. Fig. 2.15 shows the most frequently used types of display formats. In the reflective dynamic scattering display the electrode surfaces are treated so that the nematic molecules at each surface align with their long axis perpendicular to the electrode. The other molecules in the display then also follow this orientation. Incident light will now pass straight through the liquid crystal, will be reflected at the back electrode, and return to the viewer. If an external electric field is applied to the electrodes it causes a random reorientation of the crystal molecules such that the incident light is scattered in all directions and the display now appears opaque. The transmissive dynamic scattering display works on the same principle as the reflective device except that the rear electrode is also transparent and an auxiliary light source is used. The light vane ensures even illumination of the display whilst preventing the viewer from seeing the source directly. The reflective dynamic scattering display gives a good contrast ratio between operated and non-operated segments and this is independent of the intensity of the ambient light. However the inclination of the incident light is important and multiple sources can cause unwanted reflections which greatly reduce the contrast ratio. The transmissive display uses a precisely controlled source and so overcomes the problem of light inclination. It is also visible in the dark, but needs an auxiliary light source.

One form of field effect display is twisted nematic. In this the nematic molecules near the two electrodes are made to align themselves at right angles to each other so that in between there is a progressive change in orientation. Polarisers are used in the front and the back of the display and can be arranged for vertical or horizontal polarisation. Suppose the front polariser is vertically polarised and the back is horizontally polarised for a reflective display. With no power applied to the display the light will pass the first polariser and be twisted such that it goes through the second polariser and emerges after reflection. When power is applied the liquid crystal material rotates without scattering such that the vertically polarised light is not twisted during its passage through the liquid crystal and is therefore absorbed by the rear polariser. The display therefore gives dark segments on a white background. This also applies for transmissive displays. If both front and back polarisers are of the same type then one would have light segments on a dark background. It is important that the nematic molecules of a field effect display rotate and do not scatter. This means that the nematic liquid must have a high resistivity and a positive dielectric anisotropy. The response time of the display is improved by adding cholesteric to the nematic liquid so that the mixture has a natural helical structure, but too much cholesteric must not be added or it gives rotations in excess of 90°. Field effect displays have better contrast than dynamic scattering types and this is virtually unaffected by the inclination or number of ambient light sources.

Liquid crystal displays consist of two glass plates separated by an edge washer 10 μm to 20 μm thick, and the intervening space filled with liquid crystal. The inner surfaces of the glass are coated with indium–tin shaped to the required conductor pattern. The spacer seals the volume between the glass and it must be effective in preventing contamination from reaching the liquid crystal material, which would reduce the life of the display. The seal must cater for the unequal thermal expansion of the glass and the liquid crystal. Many types of sealing materials are used, the most common is solder glass since it is inert and will not contaminate the liquid. However, liquid crystals cannot usually withstand the high temperatures involved in soldering so that it is usual to leave a hole in the spacer, through which the liquid can be introduced after sealing, and to fill this hole with an inert plastic seal.

The advantage of liquid crystal displays is that they can be made in large sizes, to almost any format, at low cost. The power consumed by the display is very low; of the order of $10^{-4}$ to $10^{-5}$ W/cm$^2$ of display area for dynamic scattering and $10^{-6}$ to $10^{-7}$ W/cm$^2$ for field effect. Both types of display begin to switch at a certain voltage level called their threshold voltage. However the switching is not complete until a much higher voltage level is reached. For dynamic scattering these two levels are about 4 V and 15 V whereas for field effect they are about 1.5 V and 6 V. Field effect displays also take $\frac{1}{10}$th to $\frac{1}{100}$th the current required by dynamic scattering displays and are therefore preferred for applications such as electronic wrist watches.

Liquid crystal displays have a limited operating temperature range. At high temperatures the material will be isotropic and at too low a temperature it reverts to the solid crystalline state. These states are reversible and generally the display will not suffer any damage. Most liquid crystals can work between $-10\,°C$ and $+100\,°C$ but since the temperature affects other parameters such as speed, threshold voltage and chemical stability, the operating range of most commercial devices is $0\,°C$ to $+60\,°C$.

The operating speed of the liquid crystal display is very slow compared to a LED display since molecular rather than electron motion is involved. The turn on and turn off times are given by

$$t_{ON} \propto \eta/\Delta E V^2 \qquad (2.5)$$

$$t_{OFF} \propto \eta d^2 \qquad (2.6)$$

where $\eta$ is the viscosity, $\Delta E$ the dielectric anisotropy of the liquid crystal material, $V$ the applied voltage and $d$ the thickness of the liquid crystal layer. The operating speed is therefore a function of temperature, frequency, applied voltage, cell dimensions and the composition of the liquid crystal material. Generally the turn on time is between 2 ms and 10 ms and the turn off time 50 to 300 ms. At low temperatures below about $10\,°C$ the turn off time becomes so long that it can cause the display to appear smeared.

The drive technique for a liquid crystal display is determined by the need for alternating rather than direct voltages. It can be obtained relatively easily in logic circuits by applying antiphase waveforms to the two electrodes.

The frequency range is usually 25 Hz to 1 kHz, although for field effect displays it can be up to 10 kHz. Liquid crystal displays are also difficult to multiplex so that, for multicharacter displays, more drive electronics is needed. However due to their low power requirements the displays can be directly interfaced to MOS and their low current requirement enables high resistive connectors, such as conductive elastomers, to be used.

Filters and coloured lights can be used with liquid crystal displays, and dyes can also be added to the dynamic scattering displays to give a coloured rather than a milky display. Alternatively a thin birefringent layer, such as cellophane, can be inserted between the cell and one of the polarisers in dynamic scattering displays to create interference colours. None of these techniques are as yet widely used.

### 2.6.3 Gas discharge displays

Many different types of construction are used in cold cathode gas discharge displays. All of them employ an envelope containing neon gas across which a potential is applied. There are two processes which cause emission of photons in these devices. The first is the excitation of atoms of the gas by electron bombardment and the second is radiative recombination of ions and electrons. The light glow comes from a region close to the cathode surface so that it can be shaped to the outline of the cathode.

When a voltage is applied and increased across a neon cell the current build up is initially very slow and is due to ionisation in the gas and secondary emission processes at the cathode. The current is usually below 1 $\mu$A which is too low to give a visible glow at the cathode, with a good contrast between ON and OFF segments. When the voltage reaches a value known as the ignition threshold the current through the cell rapidly increases, which produces space charges and reduces the voltage drop until a minimum value is reached. The device now operates in a constant voltage mode and needs an external system to control its current. If the voltage is reduced below the *extinction threshold* level, the neon cell will go off and cease to glow.

Gas discharge displays give a high light output which is adequate for day-light viewing. The brightness is proportional to the current and is limited only by the power dissipation in the display since no saturation effects occur. This means that the displays can be pulsed for multiplex operation, although the shorter the pulse duration the higher the voltage needed to start a discharge. The disadvantage of gas discharge displays is that they need a relatively high operating voltage, between 150 V to 200 V, and suffer from cathode sputtering damage due to bombardment of the cathode by ions. This bombardment changes the cathode surface, therefore affecting the threshold voltages, and deposits the cathode material onto the glass envelope. The effect is worst at low gas pressures and high current densities but it can be reduced by several techniques such as adding a small amount of mercury to the neon gas. A further disadvantage of gas discharge displays is that they need a relatively long time, of the order of tens of microseconds, for ignition and for decay.

Gas discharge displays can be classed as a.c. or d.c. and within each class there is a variety of constructions. The best known and longest established d.c. display is the numerical indicator tube or Nixi. It has a single anode and several cathodes which are shaped into the form of numerals and stacked behind each other. When a discharge starts the gas molecules collect

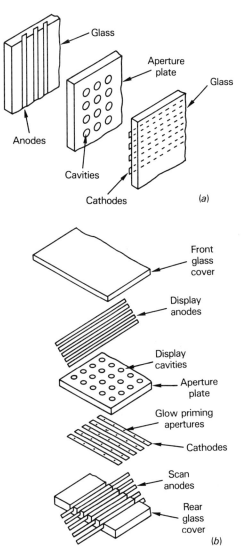

Fig. 2.16. Direct current gas discharge panel; (a) X-Y addressed panel, (b) self-scan panel.

around the cathode which is *addressed* making it glow, and making it seem larger in size than other cathodes so that it is not hidden. Alternatively one can have a planar arrangement using seven or more bars in one plane which are lit in combination to produce numerals and alphabetical characters. The colour of the display in all these types is in the 580 nm to 670 nm wavelength range (yellow to red).

Larger displays can also be constructed using the dot matrix format, either for single or multiple characters. These are also known as d.c. plasma panels and two different types are shown in fig. 2.16. In the $X-Y$ addressed panel the gas in each display cavity or cell is lit by applying a voltage across the appropriate anode and cathode. In the non-storage operating mode each cell is biased below the extinction threshold. Pulses of voltage are applied to the cathodes and anodes. The magnitude of these pulses is such that only at their coincident point, where the anode and cathode electrodes cross, will their combined voltage exceed the ignition threshold and turn on the display cell. Once the pulses disappear the cell goes out. For the storage mode of operation the cells are biased between the extinction and ignition thresholds so that after the turn on pulses are removed a lit cell remains on until turned off by a negative coincident pulse which drives its voltage below the extinction threshold.

In all d.c. panels a cell once lit operates in a constant voltage mode and its current must be limited by external means such as series resistors. Dot matrix panels need many resistors and several panels have been built which include these at the crossover locations either as thick or thin film resistors. Another disadvantage of the $X-Y$ addressed panel is that it needs a large amount of associated drive circuitry. If there are $m \times n$ cells then $n + m$ switches are needed for the address lines. The self-scan design (fig. 2.16*b*) simplifies the drive requirements at the expense of a more complex panel construction. It really consists of two separate display systems. The first is made up of the scan anodes and the rear of the cathodes forming the scan section. The second is the display anodes and the front of the cathodes giving the display section. All the scan anodes are connected via resistors to a fixed high voltage and apart from the first cathode, called the reset cathode, every third cathode is connected together and the system supplied with a three-phase clock. The display anodes are supplied with the address information so that now an $m \times n$ display only requires $m + 3$ switches. In operation a glow is first established under the reset anode and it then primes the space under the adjacent cathode so that, under the influence of the three-phase clock, a glow is constantly scanning along the back of the cathodes. The priming apertures are too small to enable this glow to be seen from the front but they prime the region in front of them so that when a voltage is applied on a display anode only the primed cell in the row will light. Once a cathode lights it draws current which causes a drop in the load resistances and extinguishes a previously lit cell so that these panels do not operate in a storage mode.

One form of a.c. panel is shown in fig. 2.17. The $X-Y$ conductors are deposited on a thick glass backing plate and are then covered with a thin film of glass. The two glass plates are separated by a spacer and this region filled with the display gas. No aperture plate is required since the discharge is usually well defined in the region of the conductor cross-overs. The gas is typically a mixture of 96% neon and 4% argon. The argon quenches neon metastables and allows rapid decay of the discharge. The mixture gives a lower operating voltage than pure neon and the voltage can be reduced still further by coating the dielectric layer with a low work function material such as magnesium oxide (MgO), which also protects the glass from sputtering.

The operation of the a.c. panel can be followed by reference to the cell voltage waveforms. A maintaining voltage, which is less than the ignition threshold $V_s$, is applied across each cell. Each intersection of the $X$ and $Y$ electrodes forms a discharge cell in which the conductors are capacitively coupled to the gas. When a voltage pulse is applied to the conductors the voltage at their intersection exceeds $V_s$ and the cell ignites. Electrons and ions separate out and deposit onto the glass giving a wall charge voltage which opposes the applied voltage and turns off the discharge in about 0.1 $\mu$s. Therefore with an alternating voltage applied to the panel the cells are discharged every half cycle.

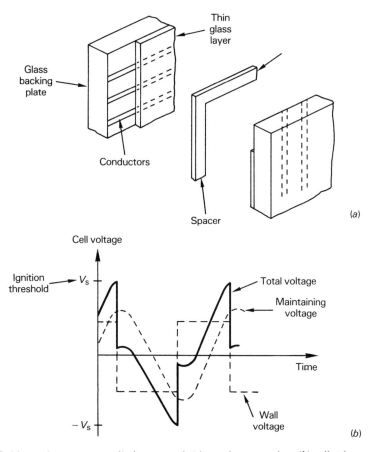

Fig. 2.17. Alternating current gas discharge panel; (a) panel construction, (b) cell voltage waveforms.

Provided the operating frequency is high enough (about 1 kHz) there should be no flicker when the panel is viewed. The panel has memory due to the wall charge since it adds to the maintaining voltage every half cycle and ensures that the total voltage exceeds $V_s$. To turn off an on cell the wall charge must be removed. This is done by applying negative pulses on the $X$-$Y$ lines to reduce the slope of the maintaining voltage at the point at which the discharge occurs. This reduces the discharge intensity to a value at which the wall charge falls to zero instead of reversing. An alternative technique is to interrupt the maintaining voltage so that the whole panel is erased simultaneously. Although sine waves are shown in fig. 2.17 a square maintaining waveform is preferred since it is easier to generate and allows greater tolerance on the voltage and timing pulses. In this system, an extinction pulse is applied to produce a discharge which reduces the wall charge to zero before the start of the opposite half cycle, causing the lit cell to go off during that half cycle. The brightness of the panel is proportional to the frequency of the voltage and its pulse amplitude. The maximum frequency is limited to about 100 kHz, above which the panel will not work due to cancellation of wall charge by residual ions when the voltage is reversed.

### 2.6.4 Other display technologies

The displays described in this section are not very widely used, either because they are relatively new devices or because they are being replaced by a newer technology. The only exception to this is the cathode ray tube (CRT) which is by far the most popularly used device for large displays. This type of display is not covered in this book and the reader is directed to reference 11 in the bibliography. However the CRT is strongly challenged by other technologies, such as plasma panel displays. It is at this moment going through several developments, such as the use of a large area electron source and thin aperture control plates, in order to reduce its bulk and so fight off this challenge.

The CRT works on the principle of fluorescence since an electron beam strikes the phosphor on the screen and produces the image. Smaller fluorescent displays are also in use. In these displays a cathode is heated to produce electrons and anodes are placed in front of it, often in a 7-bar format and coated with phosphor. A voltage applied between cathode and selected anodes causes the electrons to bombard the anode and emit light. A voltage of between 20 and 30 V is usually required with a current of about 0.5 mA per segment.

Perhaps the most traditional displays are of the incandescent type. These are available in many forms. Tiny filament lamps may be placed behind slots arranged in 7-bar format and selected lamps are lit to generate the characters. Fibre optic guides are often used to concentrate this light into small areas and so produce a very bright display. Back projection displays are also available in which any message can be selected from a film. The direct view filament display uses seven filaments arranged in a 7-bar format and these are lit to produce the display.

Incandescent displays usually operate from a 5 V supply and are capable of producing high brightness. However, increased light output is obtained at the expense of operating life and requires relatively large power input. The output covers a wide frequency spectrum and can be filtered to give any desired colour.

Electroluminescent or light emitting film displays (LEF) work on the principle that certain solids emit light when an electric field is applied across them. Usually a thin layer of phosphor is sandwiched between two electrodes and forms the emitting layer. Both a.c. and d.c. fields may be used. Alternating current displays work in the range 50 V to 300 V and 50 Hz to 5 kHz, and emit green, blue or yellow light depending on the phosphor material. The panel brightness increases with voltage and frequency but this also reduces its operating life. Direct current panels generally work at about 100 V but can be designed for 10 V to 20 V at the expense of efficiency and life. Current densities are in the order of $0.5\,\text{mA/cm}^2$ to $2\,\text{mA/cm}^2$. Electroluminescent displays have typical rise times of $20\,\mu\text{s}$ to $30\,\mu\text{s}$ and decay times of 2 ms to 3 ms.

Two displays which are currently under development are the electrochromic and electrophoretic cells. The electrochromic display works on the principle that an electric field changes the light absorption properties of certain materials. The material is initially colourless but a field colours it and this colouration remains even after the field is removed. To return to the colourless state the field must be reversed. Both transmissive and reflective displays have been built using this principle. In one type of cell a thin film of an electrochromic material, such as tungsten oxide, is evaporated onto a transparent conductor. This is then covered by a layer of insulating material followed by a second transparent conductor. When a direct voltage is applied to the two conductors, with the negative potential next to the electrochromic layer, electron injection occurs into the material turning it deep blue. When the voltage is reversed the injected electrons are extracted back to the anode and the display turns colourless again.

Electrophoretic displays have a similar construction to liquid crystal cells but work on an entirely different principle. An organic liquid containing dyes and a suspension of charged pigment particles is sandwiched between two electrodes, the front one being transparent. When a d.c. field is applied between the electrodes, the charged particles move, due to electrophoresis, towards the cathode or anode depending on the polarity of their charge. This changes the colour of the suspension as viewed through the transparent electrode. The colour of the charged particles and the liquid can be varied by dyes to give a combination of display colours. The display also has memory since once the particles have been deposited onto the surface of an electrode they remain there, even after the applied voltage has been removed, due to the attractive force between pigment particles and electrodes. The electric field must be reversed to return

## 2.7 Optical communication

| Parameter | Type of display | | | |
|---|---|---|---|---|
| | LED | Gas discharge | Liquid crystal | Incandescent |
| Colour | Red Yellow Orange Green | Neon | Clear/opaque | White (filtered to any colour) |
| Power/digit | 0.02 W to 1.0 W | 0.5 W to 1.0 W | 1 $\mu$W to 100 $\mu$W | 0.25 W to 1.0 W |
| Voltage (V) | 1.7 to 5.0 | 150 to 300 | 3 to 50 | 1 to 5 |
| Switching speed | 1 $\mu$s | 1 ms | 100 ms to 300 ms | 100 $\mu$s |
| Life (hours) | 100 000 to $\infty$ | 100 000 to 200 000 | 10 000 to 50 000 | 50 000 to 250 000 |
| Brightness | Good | Good | Not applicable | Excellent |
| Large panel capability | No | Yes | Yes | No |

Fig. 2.18. Comparison of a few commonly used display technologies.

the display to its original state. The electrophoretic display is relatively slow having rise and fall times of the order of 20 ms and 100 ms respectively.

Figure 2.18 compares the properties of a few commonly used display technologies.

### 2.7 Optical communication

In recent years optical technology has been used for telecommunications and digital data processing. This has largely been due to the development of commercial fibre optical cables with attenuation below 20 db/km and with bandwidths exceeding 1 GHz/km. The advantages of optical communication over methods using conventional electrical cabling are the following:

(i) A dielectric waveguide is used so that the system does not radiate radio frequency interference (RFI) and is not affected by externally generated interference.

(ii) No return line is required which means that interconnection between systems is easier since they are not referenced to a ground potential.

(iii) Glass, used in optical fibres, is inherently a cheaper material than copper.

(iv) Optical fibres are typically less than 0.5 mm in diameter which is about $\frac{1}{20}$th the size of a conventional coaxial copper cable. Therefore many optical fibres can be grouped together into a single cable without making it too bulky.

An optical communication system uses three basic components, a light source, optical cable and a light detector. Light emitting diodes and injection lasers are used as sources since they have high reliability and long life, and can be directly modulated because their light output is proportional to the input current. A laser is generally preferred due to its ability to produce collimated light with a spectral width less than 2 nm, and a small divergence angle which gives efficient coupling between the laser and the fibre optic cable. The laser also has a response speed in the sub-nanosecond region, after it has reached its lasing threshold, and a high slope efficiency in the region of 25%, measured in terms of megawatts of optical power output per milliamp of input drive. At present, lasers have a shorter life than LEDs.

The fibre used in optical communication must have a low loss. The theoretical limit is a loss of 1 db/km at which molecular scattering starts to occur, although commercial fibres are closer to 3 db/km. The attenuation in the cable is constant over the frequency band, unlike coaxial cables in which the loss increases with frequency. It is above about 10 MHz that the loss in the optical cable falls below that of a coaxial cable. The attenuation in optical fibres increases at low wavelengths due to scattering. However it is relatively constant at

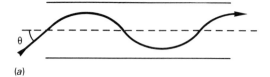

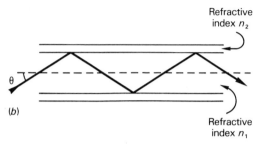

Fig. 2.19. Fibre optic waveguides; (a) graded index (refraction), (b) step index (reflection).

the 3 db/km value over the 800 nm to 900 nm range, which is most frequently used with LEDs and lasers.

Many types of construction have been used for fibre optics. Generally they are about 100 μm to 200 μm in diameter and the refractive index of the glass is greater at the centre than at the periphery. This change can be gradual or steep as shown in fig. 2.19. In the graded fibre the light is continuously bent towards the axis of the fibre so that it works by refraction. In the step fibre light propagates by total reflection at the interface of the two materials and there is no refraction. Generally, the graded index cable gives a higher data rate than the step index fibre.

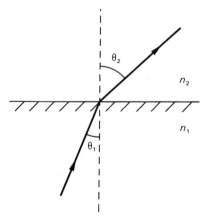

Fig. 2.20. Illustration of Snell's law, $n_1 \sin \theta_1 = n_2 \sin \theta_2$.

The operation of the step index fibre can be explained by Snell's law (fig. 2.20) given by

$$n_1 \sin \theta_1 = n_2 \sin \theta_2$$

When light enters the boundary between two transparent media it is partially reflected and partially refracted. In order to have total reflection $\theta_2$ must be 90° or greater so that the critical angle, $\theta_1$, when internal reflections start is given by

$$\sin \theta_1 = n_2/n_1 \qquad (2.7)$$

From this it can be shown that the maximum angle $\phi$ at which the ray can enter the waveguide and be totally reflected is given by

$$\sin \phi = (n_1^2 - n_2^2)^{1/2} \qquad (2.8)$$

This angle is called the numerical aperture of the fibre.

Waveguides are bundled together in groups of 60 to 1000 fibres to form cables. The bundles are usually sheathed in PVC and their ends are made optically flat in order to provide an efficient transfer for input and output light. Several optical cables can be joined together using special low loss connections which give losses of typically 2 db to 5 db. For light transmission the position of the fibres is not critical although this is clearly important if a picture is to be transmitted.

Avalanche photodiodes (APD) are almost exclusively used as photodetectors in optical transmission systems. Since most of the noise in such a system is associated with the receiver an avalanche photodiode is preferred since its multiplication effect gives it a higher signal to noise ratio.

A typical optical system uses lasers, modulators, detectors, lenses and so on. Considerable effort has gone into developing techniques which enable these parts to be integrated onto a single substrate. This has resulted in an optical analogy of an integrated circuit and it is called integrated optics. Both hybrid and monolithic systems have been built. In a hybrid circuit different technologies are used for the various parts of the system. A monolithic circuit uses materials which optimise the performance of some sections but not all of them. The advantage of a monolithic circuit is ease of manufacture and small size. Integrated optics is still very much in the laboratory stage but shows advantages over discrete optical communication systems for modulation frequencies in excess of 300 MHz.

## 2.8 Holography

Holography is a technique for creating three-dimensional images. The word holography is derived from the Greek words *holos* meaning entire and *graphein* meaning recording. Whereas only the amplitude of a light wave from an object is recorded in conventional photography, in holography both the amplitude and phase of this wave is recorded. These are stored in interference patterns formed between the object wave and a reference wave. The interfering waveforms must be coherent so a laser source is commonly used. Fig. 2.21a shows a system for recording holograms. The light from the laser is split, one beam going to the object and then onto the film whereas the second beam goes direct to the film after reflection from a mirror. The hologram is formed as an interference pattern on the film. When this is illuminated by the reference beam alone light passing through the plate is selectively transmitted or absorbed so as to create the original object wave. The observer sees a virtual image which looks as if the original object was in place, including three-dimensional effects which result as the viewing angle is changed.

There are many different types of holograms. Fig. 2.21 shows the system used to record and reconstruct a Fresnel hologram. In this system the object is a finite distance from the recording medium. If the object is effectively at an infinite distance Fraunhofer holograms are produced; a special case of this is the Fourier transform hologram in which the photographic film is placed at the focus of a lens and stores the Fourier transform of the object. If the photographic film is relatively thick then the interference patterns are recorded thoughout the depth of the recording media. These are called volume or thick holograms and, since they are more selective in both wavelength and direction than plane or thin holograms, they have a higher storage capacity. If the reference and object waves are introduced from opposite sides of the hologram then white light can be used to reconstruct the image. The discussions so far have concentrated on the absorption type of hologram in which the amplitude of the transmittance of the hologram varies. An alternative type of hologram, called the phase hologram, varies the phase of the transmitted wave. These are generally more efficient than the absorption types.

Holography has many applications, two of which are discussed here. The ability to produce three-dimensional images is clearly very attractive in display systems. In particular holography has been applied to microscopy where a recording of a moving object can be made and then later reconstructed and examined in detail including focusing at different depths. Holograms are also capable of storing large amounts of data. Since the stored information is spread over the whole area of the storage medium a defect in one part of the area still gives the complete image although it reduces its signal to noise ratio. Holographic storage is mainly used in read only memories although work is being done on the development of reversible media for read–write stores.

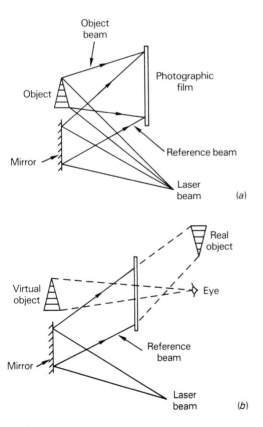

Fig. 2.21. Holograms; (a) recording, (b) reconstructing.

# 3. Resistors

## 3.1 Introduction

A resistor is probably the simplest and the most commonly used electronic component. It is partly due to this that many designers consider resistors to require very little thought so that the component is often incorrectly applied. In this chapter the construction and characteristics of the various types of resistors will be described. They fall into three main groups, linear fixed, linear variable and non-linear.

## 3.2 Resistor principles

### 3.2.1 Equivalent circuit

Ohm's law is well known and commonly stated as

$$R = V/I \qquad (3.1)$$

where $R$ is the resistance of the circuit, $V$ the voltage across the circuit and $I$ the current flowing through it. The inverse of resistance, $1/R$ is called conductance. The unit of resistance is the ohm ($\Omega$) and the unit of conductance is the siemen or mho ($\Omega^{-1}$).

Equation (3.1) represents a simplified picture of an actual resistor.

A practical resistor is a complex component whose *equivalent circuit* is similar to that shown in fig. 3.1a. The inductance ($L$) is that of leads and terminations and the capacitance ($C$) is distributed over the body of the resistor. The magnitudes of $L$ and $C$ vary with resistor composition, resistance value and operating frequency. Therefore the variation of resistor impedance with frequency can be fairly complex. The frequency characteristic for one type of resistor is shown in fig. 3.1b.

The impedance of high value resistors decreases with frequency whilst that of low value resistors increases. Errors occur even at low frequencies if it is assumed that the value of a resistor is given by (3.1). For low value resistors these errors are caused by the resistance of the leads, and the contacts between the leads and the resistor body. For high value resistors they are due to parallel leakage paths across the body of the resistor, such as through any protective material used around the resistor element.

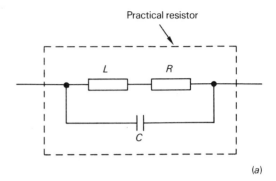

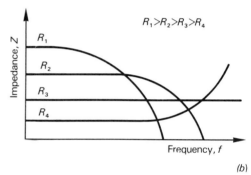

Fig. 3.1. A typical resistor; (a) equivalent circuit, (b) frequency characteristics.

### 3.2.2 Resistor ratings

The maximum power dissipation of the resistor is probably one of its most important ratings. It is measured in watts (W) and is given by

$$W = VI \qquad (3.2)$$

or

$$W = V^2/R \qquad (3.3)$$

The peak voltage and current ratings are often

## 3.2 Resistor principles

determined by the need to limit the resistor dissipation, as given by (3.2) and (3.3). The maximum power rating is determined by the ability of the resistor to dissipate internally generated heat. The higher the ambient temperature the lower the allowable internal dissipation, and data sheets give power *derating* curves for resistors operating at different ambient temperatures. If a resistor is operated beyond its rated power it can fail due to an open circuit, a short circuit, or burst into flames.

The maximum voltage rating of a high value resistor, or a resistor operated in a pulsed mode, is often limited by considerations other than peak power dissipation. At these voltages, high electric fields are set up in the resistor body which can cause irreversible changes in resistance value, or lead to dielectric breakdown.

### 3.2.3 *Resistor characteristics*

Many characteristics need to be considered when choosing a resistor. The selection tolerance represents the band around the specified value within which the manufacturer guarantees his resistor will fall. Common tolerances are ± 10%, ± 5%, and ± 1%. Commercial resistors are available in a series of values such that one can cover the whole range of resistances in a given band. 10% tolerance resistors are available in the E12 series which are (in units of ohms)

1.0  1.2  1.5  1.8  2.2  2.7  3.3  3.9  4.7
5.6  6.8  8.2

This means that, for example, the 1.8 Ω resistor will cover a range up to 1.8 Ω + 10% i.e. 1.98 Ω and the 2.2 Ω resistor will cover a range down to 2.2 Ω - 10% i.e. 1.98 Ω so that any value between 1.8 Ω and 2.2 Ω can be obtained by selection from the manufactured range which would be distributed around the nominal value. The E24 series covers 5% tolerance resistors and these are available in the following nominal values (ohms):

1.0  1.1  1.2  1.3  1.5  1.6  1.8  2.0  2.2
2.4  2.7  3.0  3.3  3.6  3.9  4.3  4.7  5.1
5.6  6.2  6.8  7.5  8.2  9.1

The E96 series covers 1% resistors and these are available in the following values (ohms):

1.00  1.02  1.05  1.07  1.10  1.13  1.15
1.18  1.21  1.24  1.27  1.30  1.33  1.37
1.40  1.43  1.47  1.50  1.54  1.58  1.62
1.65  1.69  1.74  1.78  1.82  1.87  1.91
1.96  2.00  2.05  2.10  2.15  2.21  2.26
2.32  2.37  2.43  2.49  2.55  2.61  2.67
2.74  2.80  2.87  2.94  3.01  3.09  3.16
3.24  3.32  3.40  3.48  3.57  3.65  3.74
3.83  3.92  4.02  4.12  4.22  4.32  4.42
4.53  4.64  4.75  4.87  4.99  5.11  5.23
5.36  5.49  5.62  5.76  5.90  6.04  6.19
6.34  6.49  6.65  6.81  6.98  7.15  7.32
7.50  7.68  7.87  8.06  8.25  8.45  8.66
8.87  9.09  9.31  9.53  9.76

Many factors may affect the operating value of a resistor. The most common factor is temperature change. The temperature coefficient of resistance (TCR) of a resistor, expressed either as a percentage change in resistance per degree centigrade or parts per million (ppm) change per degree centigrade, is used to quantify this variation. A *positive temperature coefficient* means that resistance increases with temperature and a *negative coefficient* indicates a decrease.

In some types of resistors the resistance varies with the magnitude of the applied voltage, and this can be quantified as percentage change in resistance per volt. This is known as the *voltage coefficient* of resistance.

All resistors also exhibit a change in resistance over a period of time. This is referred to as the *long term stability* of the resistor and is primarily caused by heating effects within the body of the resistor. The long term stability can be greatly improved by running the resistor at below its rated power so that its body temperature is comparatively lower. Resistor data sheets often provide the designer with graphs, such as those shown in fig. 3.2, which show the relationship between power dissipation and long term stability. For example, if this resistor is operated in an ambient temperature of 40 °C with 0.5 W dissipation then it will run at a body temperature of 120 °C (fig. 3.2*a*) and a 1 MΩ resistor will drift by a maximum of 3% after 1000 hours operation (fig. 3.2*b*). If the same resistor is run with 0.25 W dissipation then it will attain a peak body temperature of 80 °C and a maximum drift of below 1.5% after 1000 hours.

Another characteristic of importance to the circuit designer is resistor noise. When a voltage is applied across a resistor it results in an electron drift current within the body of the resistor. This current is proportional to the applied

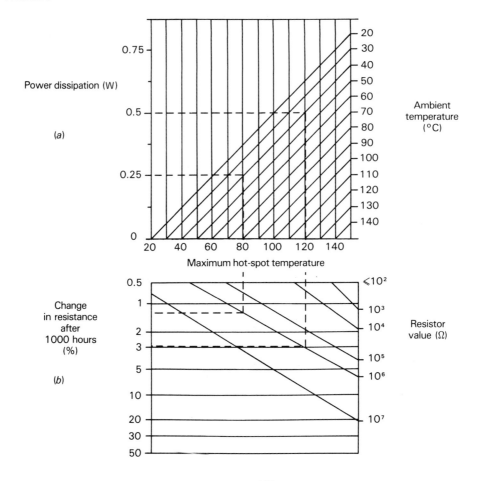

Fig. 3.2. Resistor stability curves.

voltage and represents the desired signal. In addition, a random motion of free electrons and a thermal motion of molecules within the resistor is set up and this causes a noise current. It occurs in all types of resistors and is known as thermal or Johnson noise. Some resistors also exhibit another type of noise called current noise. This is due to the random changes in the composition of the resistor material caused by current flowing through it. Resistor noise is measured in either units of microvolt of noise per volt of applied potential, or in decibels using a fixed reference such as $1\mu V/V$. The noise voltage may be reduced by operating a resistor below its rated power dissipation, as required for improved stability, but this often results in a physically larger and more expensive resistor.

Several mechanical and environmental parameters need to be considered when selecting resistors. The ambient temperature and humidity of the intended place of operation often determines the choice of external protection applied to the resistor body. Printed circuit board assembly techniques determine the maximum lead bending stress the resistor must stand. Apart from causing damage to the leads this stress can result in an increase in resistance value, outside the original tolerance of the resistor. The solderability of the resistor is specified as a maximum change in resistance value when it is being soldered into a board, the solder bath temperature and the duration of the soldering operation being specified. Since board cleaning agents and solder flux are often likely to come into contact with a resistor, its resistance to chemicals is another factor which the design engineer often has to consider.

## 3.3 Fixed linear resistors

It is clear from the previous section that the end of life value of a resistor is very different from its nominal value. The nominal value drifts at several stages including initial selection, printed circuit board assembly, storage and operation. The amount of drift in each one of these stages is determined by the type of resistor, so the maximum allowable resistance change often determines the type of resistor to be used in a given application. In this section carbon composition, carbon film, metal oxide film, metal film, metal glaze and wirewound resistors are described.

### 3.3.1 *Carbon composition resistors*

Carbon composition resistors are probably the best known and most widely used although they are now becoming obsolete. The resistive element consists of finely ground carbon particles mixed with refractory filling. These are bonded together using a synthetic resin binder and the mixture is then either compressed into shape and fired in a kiln or coated onto a substrate such as glass and then fired. The composition of the mixture determines the resistance value. Many techniques may then be used for the final construction. For example, metal end caps with leads can be forced onto the carbon rod, or the ends of the rod can be sprayed with metal and the leads soldered to it. Both these techniques give a non-insulated construction. For an insulated resistor, the ends of the rod may be sprayed with metal and metal caps fitted over them. The resistor body is then encapsulated in a *thermoplastic* insulation, in a ceramic tube, or by coating it with a moisture resistant silicone lacquer. Insulated resistors are more commonly used on a printed circuit board although they have a lower power dissipation capability than non-insulated types and are therefore larger for the same power ratings.

Carbon composition resistors are rugged and will withstand large surges of current. However they have large voltage and temperature coefficients of resistance, poor stability and exhibit high noise, especially current noise.

### 3.3.2 *Carbon film resistors*

In a carbon film resistor, a thin layer of carbon film is formed onto a ceramic rod. This is done by heating the rods and tumbling them in a retort in the presence of hydrocarbon gas. At about 1000 °C the gas decomposes or 'cracks' to deposit the carbon film onto the rods. The thickness of the film is determined by the gas concentration and the exposure time. End caps are now pressed over the carbon film or the ends are metallised and leads soldered to them. The resistance value of the film can be increased by a factor of 10 to 10 000 by cutting the deposited film or 'blank' in a spiral. A diamond cutting wheel or a laser can be used for this. The laser gives cleaner edges so that the performance of the resistor is better. To obtain larger resistor values, the pitch of the spiral can be reduced. The stability and reliability can be improved by increasing the thickness and decreasing the resistivity of the film layer.

To ensure uniform heat flow along the resistor body the helix must be evenly distributed and this is automatically adjusted during production. A laser cutting machine has a throughput of about 20 000 resistors per hour. The carbon film is delicate and after cutting it is protected by enclosing it in a ceramic or glass tube or by applying several coats of varnish to it.

The noise produced by a carbon film resistor is much less than that of a carbon composition type. It is proportional to the voltage stress in the film, and inversely proportional to both the film thickness and the square of the length of the film. The temperature coefficient of resistance is a function of the amount of effective carbon film used and increases significantly at high values of resistance. Carbon film resistors tend to fail due to irregularities in the film track which result in hot spots and noise, especially at high resistor values.

### 3.3.3 *Metal oxide resistors*

The film material used in metal oxide film resistors is a hard glassy oxide formed on a glass or ceramic rod. Many techniques are used in the manufacture of the resistor. In one method, tin oxide, which is a semiconductor material, is doped with antimony and is used as the resistive element. The antimony is used to control the electrical characteristics of the film. In the manufacturing process, molten glass is extruded at about 1000 °C and a tin–antimony chloride solution is sprayed onto it. The solution decomposes and solidifies to form a coating of oxide on the glass surface. After cooling, the glass is cut into individual resistors, the ends are metallised, spiralled, capped and encapsulated by processes similar to those used for carbon film resistors. The film used is generally between $10^{-9}$ and $10^{-7}$ m thick and has a resistivity of about $10^3$ $\Omega$ per square and a temperature coefficient of resistance (TCR) of $\pm 250$ ppm/°C.

The TCR and noise of metal oxide film resistors are lower than those for carbon film and vary only slightly with the resistance value. Although the TCR of carbon film resistors is negative it can be either positive or negative for metal oxide resistors because metal oxide is a semiconductor material.

### 3.3.4 Metal film resistors

Metal film resistors are formed either by vacuum evaporation of nickel–chromium alloys onto cylindrical ceramic substrates, or by chemical deposition of nickel alloys onto the substrates. The first process gives a film having a surface resistivity of about $10^3$ $\Omega$ per square and the second gives one of $10^5$ $\Omega$ per square. In both cases the quality of the resistor depends largely on the control exercised over the process. After the film has been formed it is spiralled using a similar process to that used for carbon film resistors. Since the metal film is attacked by humidity it is protected using several techniques, e.g. by lacquering and encapsulating in a moulded plastic case, by sealing in a resin filled tube, or by lacquering and *hermetic sealing* in a ceramic tube.

Metal film resistors have good stability and a very low TCR which may be positive or negative. They are therefore widely used as general purpose and semi-precision resistors.

### 3.3.5 Metal glaze resistors

The resistive material used in metal glaze resistors consists of a suspension of metal and glass particles in an organic solvent. To form metal glaze resistors, an alumina or steatite rod is dipped into the glaze and withdrawn at a controlled rate, the thickness of the film adhering to the rod being inversely proportional to its withdrawal rate. The rods are then fired at 900 °C to 1100 °C which causes the glass to flow and form microencapsulations for the metal particles as well as binding them to the rod. The ends of the rod are then nickel plated and leads are soldered to it. The remaining processes, such as spiralling are as for carbon film resistors.

The metal glaze resistor uses a film of about $10^{-5}$ m in thickness and this gives it a high surge current capability. The resistivity of the film can be varied between 10 $\Omega$ per square and $10^6$ $\Omega$ per square by changing the percentage of glass to metal in the glaze. However, for very low resistance values there is an excess of metal which makes the resistor unstable, and for very high resistance values there is too much glass which results in intermittent operation. In these instances the metal chosen must have a lower or higher resistivity so that the percentage of metal by weight in the glaze remains between 25% to 75%.

Since high firing temperatures are used in the preparation of metal glaze resistors they can subsequently be run at high body temperatures so that relatively high power dissipation can occur in small body dimensions. The noise generated is intermediate between that generated in carbon film and carbon composition types. The TCR is low and may be positive or negative, and stability is good since the operating body temperature is usually well below its maximum capability. The metal glaze film is much thicker than either the carbon or metal film so that it is much more rugged and resistant to environmental extremes such as shock, vibration and overloads.

### 3.3.6 Wirewound resistors

The wirewound resistor is usually constructed by winding resistive wire onto a ceramic bobbin and fixing the ends to terminal leads. Due to the difference in expansion between the wire and bobbin this method of construction can result in stress in the wire so that a floating wire or bobbinless construction is often used. In this, the resistance wire floats in a slurry which is sealed in a metal can. The slurry anchors the wire, without straining it, and also provides the heat transfer medium to the case.

The wire material is usually nickel–chromium since it has good stability, low TCR and high operating temperature. High power resistors have their windows exposed to air although most general purpose resistors are encapsulated in vitreous enamel, lacquer or silicone cement. Vitreous enamel gives the resistor excellent environmental protection and therefore high reliability and long life. The material requires high curing temperatures and this results in a large TCR and selection tolerance. Silicone cement limits the body temperature to about 250 °C and so reduces the resistor power rating. The lower curing temperature gives good TCR and lower selection tolerance. The material is also cheaper than vitreous enamel.

Wirewound resistors generally fail (i) due to blemishes in the wire, especially for high value resistors, (ii) due to differences in expansion of the coating, usually vitreous enamel, and the ceramic bobbin which causes the coating to crack and let in moisture, (iii) due to electrolytic erosion at the junction of dissimilar metals e.g. between the winding and the end cap. The resistors have very low noise which is usually all Johnson noise. Their high frequency performance is poor. Inductive effects can be reduced by using two coils wound in antiphase so that the current flows in opposite directions in adjacent turns. However, at frequencies above about 5 MHz *stray capacitances* become important so that non-inductive windings are not very effective.

### 3.3.7 *Resistor comparisons*

Fig. 3.3 summarises the characteristics of the resistors discussed in this section. The characteristics of each type can to some extent be varied during manufacture to suit the application, and this is illustrated in fig.3.4.

## 3.4 Variable linear resistors

The resistors described in this section are constructed so that the resistance between two of the device terminals can be varied by the user. These resistors are known as potentiometers and they vary in size and construction. Large panel mounted devices are generally designed for applications where relatively frequent adjustment may be required, while small printed circuit board mounted devices, sometimes known as trimmer potentiometers, are used in applications where only a few adjustments will be made during the life of the equipment. Both types of potentiometers can be single turn, in which the wiper sweeps across the whole resistance element for one turn of the control spindle, or multiturn, in which many full rotations of the spindle are required to move the wiper across the resistance element. Multi-

| Parameter | Carbon composition | Carbon film | Metal oxide film | Metal film | Metal glaze | Wirewound |
|---|---|---|---|---|---|---|
| Resistance range | 10 Ω to 22 MΩ | 1 Ω to 10 MΩ | 10 Ω to 1 MΩ | 10 Ω to 1 MΩ | 10 Ω to 22 MΩ | 1 Ω to 1 MΩ |
| Selection tolerance (%) | 5 to 20 | 5 | 1 to 5 | 1 | 1 | 5 to 10 |
| Maximum d.c. volts | 500 | 500 | 500 | 350 | 350 | 500 |
| VCR (%/V) | 0.3 | 0.01 | Negligible | Negligible | Negligible | Negligible |
| TCR (%/°C) | 0.1 | 0.025 to 0.1 | 0.005 to 0.025 | 0.01 | 0.01 | 0.01 |
| Stability after 1000 hours at 70 °C (%) | 20 | 1 | 0.5 | 0.1 | 0.5 | 1 |
| Noise (μV/V) | 2 + log (R/1000) | 0.1 | 0.05 | 0.01 | 0.5 | Negligible |
| Frequency range (MHz) | 10 | 50 | 50 | 50 | 50 | 5 |
| Typical maximum power (W) | 2 | 2 | 10 | 1 | 2 | 50 |

Fig. 3.3. Summary of resistor characteristics.

| Category | Type | Resistance range | TCR (%/°C) | Power (W) | Load stability after 1000 hours at 70 °C (%) |
|---|---|---|---|---|---|
| General purpose (1) 5% tolerance TCR > 0.02%/°C | Carbon composition | 10 Ω to 22 MΩ | 0.1 | 0.25 to 2 | 20 |
| | Carbon film | 1 Ω to 10 MΩ | 0.025 to 0.1 | 0.25 to 2 | 1 |
| | Metal oxide film | 10 Ω to 1 MΩ | 0.025 | 0.25 to 2 | 2 |
| | Metal glaze | 10 Ω to 22 MΩ | 0.025 to 0.05 | 0.25 to 2 | 1 |
| General purpose (2) 1%–5% tolerance TCR < 0.02%/°C | Metal oxide film | 10 Ω to 1 MΩ | 0.01 | 0.1 to 0.5 | 0.5 |
| | Metal film | 10 Ω to 1 MΩ | 0.01 | 0.00675 to 2 | 0.2 |
| | Metal glaze | 10 Ω to 22 MΩ | 0.01 | 0.25 to 2 | 0.5 |
| Precision < 1% tolerance TCR < 0.01%/°C | Metal film | 10 Ω to 1 MΩ | 0.001 to 0.01 | 0.25 to 1 | 0.05 |
| | Wirewound | 0.1 Ω to 1 MΩ | 0.001 | 0.25 to 2 | 0.005 |
| Power > 2 W | Metal oxide film | 10 Ω to 1 MΩ | 0.05 | 10 to 200 | 5 |
| | Metal glaze | 100 Ω to 500 kΩ | 0.05 | 2 to 10 | 1 |
| | Wirewound | 1 Ω to 1 MΩ | 0.01 | 3 to 300 | 1 |
| Networks | Thick film | 10 Ω to 10 MΩ | 0.01 | 2 | 1 |
| | Thin film | 1 Ω to 50 MΩ | 0.005 | 2 | 0.5 |

Fig. 3.4. Resistor selection table by application.

turn potentiometers are capable of more accurate setting but are more expensive than single turn devices.

### 3.4.1 Potentiometer characteristics

Although many of the potentiometer characteristics such as power rating, temperature coefficient and maximum voltage are similar to those used for resistors, a potentiometer has several other unique characteristics. The end points of the potentiometer are the points on the resistive element between which the resistance varies in the required manner. These points are usually different from the end terminals. The electrical travel is the distance between these end points, where there is continuity between the wiper and the resistance element. The electrical overtravel of the potentiometer is the space beyond the electrical travel. The end resistance is the resistance between the wiper and each end terminal when the wiper is moved as far as it will go towards that terminal. The potentiometer jump-off voltage is the output voltage step, expressed as a percentage of the applied voltage, which occurs when the wiper moves from an end point onto the electrical travel. The total resistance of the potentiometer is the resistance between its end terminals.

The resistance element can often have one or more taps along its length. There are two ways in which these taps can be connected. In the first method, known as a current tap, the conductor of the tap crosses the entire width of the resistance track in a direction perpendicular to the wiper path. This tap has the capability of carrying the same magnitude of current as the end terminals but it disturbs the current distribution in the resistance element and hence the resistance law for which it was designed. The second type of tap is called a voltage or zero width tap. In this method the conductor of the tap just touches the edge of the resistance element so that it has negligible effect on the resistance law although it can carry only a fraction of the end terminal current.

The smallest incremental travel of the wiper which is needed to produce an incremental change in output voltage is called the resolution of a potentiometer. It is generally measured as a percentage of the electrical travel. Some types of potentiometers, such as the wirewound construction described later, have poor resolution since the output is stepped, the wiper having to move along the surface of the resistance element in increments of one turn. In some precision potentiometer designs this is avoided by having the wiper follow the windings spirally.

The function of a potentiometer is the relationship between its output voltage ratio and the wiper position. The amount by which the function actually follows the theoretically predicted values is called the conformity. Potentiometer linearity is a special case of conformity in which the function is a straight line, as shown in fig.3.5. There are two types of linearity. Absolute linearity is the maximum deviation of the actual characteristic from the linear characteristic, expressed as a percentage of the total applied voltage. The independent linearity is similar to the absolute linearity but now the expected line is drawn to minimise the deviations of the actual output from this line. It is usually the independent linearity which is specified on manufacturers' data sheets

Mechanical travel of the potentiometer is the total travel of the wiper between internal stops. If no stops are used then the travel is said to be continuous. The direction of travel is usually specified as viewed from the control spindle end of the potentiometer. Potentiometer shaft run out is a measure of the eccentricity of the control spindle axis with respect to its rotational axis. Radial play is the movement of the control spindle when a specified load is applied perpendicular to its axis and at a specified distance from its mounting face. End play is the movement of the spindle when load is applied along its axis. The starting and running torque are the minimum torques needed to start and keep the wiper moving in either direction. The backlash

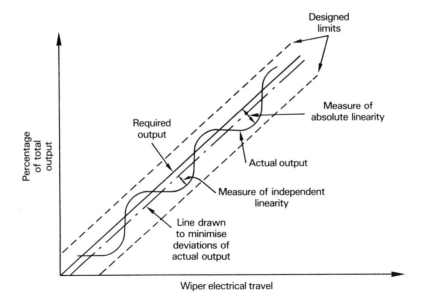

Fig. 3.5. Illustration of linearity.

is usually measured as the maximum difference in spindle position when the same output voltage ratio is set starting with the wiper at different ends of the resistance element.

Many potentiometer parameters are similar in principle to those used for resistors but there are often differences in detail. For a linear potentiometer the effect of temperature is similar to that of a conventional resistor. However, several different types of resistance elements may be combined to give a special law and in these cases the conformity can change significantly with temperature. There are several causes of noise in potentiometers:

(i) Thermal agitation or Johnson noise, which is similar to that experienced in resistors.

(ii) E.m.f. generated due to the friction of the dissimilar metals used in the wiper contact and the resistance element. This is known as the tribo-electric effect.

(iii) Change in contact resistance of the wiper. This may be due to several causes such as the wiper momentarily leaving the surface of the resistance element, or due to dirt in the wiper path. Also, if excessive current flows in the wiper circuit the resistive element can become oxidised and 'noisy'. Generally most of the noise in potentiometers is due to the change in contact resistance. It can be minimised by the use of special constructions, such as multi-fingered wipers which are spaced equidistantly across the width of the resistance element and so spreading the current and reducing the chance of lift-off caused by shock or vibration.

The resistance tolerance of a potentiometer is not as important as its linearity because the potentiometer is often used in voltage divider applications. Potentiometer life is measured in the number of times the wiper can be moved over the whole surface of the element without degrading the potentiometer characteristics. Even when the potentiometer is used over a small area of its resistance element this value of life should not be exceeded since this section of the element can wear out. A potentiometer can reach the end of its life due to a physical malfunction, or due to a drift in some parameter beyond its specified value. Generally many parameters change with usage, a few examples being total resistance, noise, linearity, torque and shaft play.

Environmental protection is required for potentiometers since the resistance element is very susceptible to attack. If the potentiometer is hermetically sealed the spindle needs to be brought out through the cover using gaskets or O-rings. This will increase the friction on the spindle and therefore the torque needed to turn the potentiometer. The torque can be reduced by lowering the wiper pressure, and this has the added advantage of increasing the life although the contact resistance may now increase.

**3.4.2** *Types of potentiometers*

Many different techniques are used for making potentiometers and these are compared in fig. 3.6.

| Parameter | Wirewound | Carbon composition | Carbon film | Metal film | Conductive plastic | Cermet |
|---|---|---|---|---|---|---|
| Resistance range | 1 Ω to 200 kΩ | 50 Ω to 100 MΩ | 100 Ω to 1 MΩ | 10 Ω to 500 kΩ | 50 Ω to 5 MΩ | 10 Ω to 5 MΩ |
| Power rating | 1 | 3 | 6 | 5 | 4 | 2 |
| Temperature coefficient | 1 | 6 | 5 | 2 | 4 | 3 |
| Linearity | 4 | 6 | 2 | 2 | 1 | 5 |
| Noise | 1 | 6 | 3 | 1 | 3 | 5 |
| Frequency range | 6 | 5 | 1 | 1 | 1 | 4 |
| Resolution | 6 | 1 | 1 | 1 | 1 | 5 |
| Life | 6 | 3 | 2 | 3 | 1 | 3 |

Fig. 3.6. Comparison of the major potentiometer technologies; 1 = best, 6 = worst.

The wirewound construction is the oldest and the most widely used for precision applications. The construction consists of a resistance wire wound onto a former, and a wiper which moves across it. This means that the potentiometer has poor resolution due to the finite steps from one turn to the next. Wirewound potentiometers have the lowest TCR of the order of $\pm 20$ ppm/°C to $\pm 100$ ppm/°C with good linearity and very low noise. The potentiometers are, however, susceptible to catastrophic failure due to breakage of a turn, have low life, and are not very useful above about 5 kHz due to the capacitive effect between turns.

Carbon composition potentiometers use an element made of moulded carbon composition similar to that used for linear fixed resistors. The resolution is now high since theoretically the resistance can be set at any value between the two extremes. The potentiometer is rugged but suffers from the disadvantage of a high TCR, poor linearity and high noise. Carbon film and metal film potentiometers are also made by similar techniques to those used for linear resistors and they overcome several of the disadvantages of carbon composition potentiometers, but are less rugged.

Conductive plastic potentiometers are made in many ways. One method consists of starting with the moulded substrate of insulating plastic and screening on the terminations and any required taps. The resistive material, consisting of thermosetting plastic containing fine carbon particles, is then screened or sprayed onto this, and the element placed in a mould under high temperature and pressure. This causes the particles to co-mould with the substrate and terminations, giving a high reliability connection since the terminations and taps are moulded integrally with the resistive track.

The conductive plastic element can be trimmed by cutting away parts of the resistive track to give any required function. The potentiometer can be made with a linearity of about $\pm 0.03\%$, resolution of better than 0.05%, long life and good frequency characteristics. If run at large currents the element may pit and degrade but catastrophic failure is rare. The conductive plastic potentiometer suffers from a high TCR so that it is more often used as a voltage divider than as a variable resistor.

Cermet potentiometers are made by similar techniques to those used for cermet resistors, where a paste consisting of fine metal particles in a glass base is applied to a ceramic former and fused at high temperature. The resulting element can dissipate high power within a small area, although it is difficult to trim to obtain non-linear functions. The wiper also wears as it rotates on the cermet element, so that the operating life is relatively short.

## 3.5 Non-linear resistors

Some resistors are referred to as non-linear since their value does not change linearly with change in one or more of the external conditions. The light dependent resistor is one type of non-linear resistor and this was described in chapter 2. *Magneto resistors*, whose values vary with the strength of a magnetic field, will be described in chapter 5. In this section, two classes of non-linear resistors will be described: (i) thermistors, whose resistance value changes with temperature; and (ii) voltage dependent resistors (VDR), whose value changes with the magnitude of the applied voltage.

The manufacturing technique used for a thermistor or a VDR is very similar. The raw materials are first ball-milled together to obtain a good mixture of the required particle size. The resulting slurry is filtered and dried and then mixed with an organic binder, and shaped by pressing into discs or extending into rods. The components are then sintered at about 1200 °C to 1500 °C. Electrical connections are made to the ends by various techniques such as metal spray, evaporation, metal paints, screen printing metal pastes and firing, and the leads are then connected. The whole device is protected by a lacquer coating or by plastic moulding. The electrical properties of the device are determined by the initial material used, the size and the manufacturing process parameters. It is important to remember that although non-linear resistors are made from semiconductor materials they do not contain any pn junctions.

### 3.5.1 *NTC thermistors*

The most common materials used for negative temperature coefficient (NTC) thermistors are the oxides of the iron group of elements (such as chromium, manganese, iron, cobalt and nickel). These materials normally exhibit a high resistivity but they are converted into semiconductors by the addition of small amounts of ions having a different valency, for example a few iron ions in iron oxide being replaced by titanium.

NTC thermistors are available in a variety of shapes and sizes and have resistance values between 0.1 Ω and 100 MΩ at 25 °C. They can be designed to operate in the range − 100 °C to + 300 °C and a single device can operate over approximately a 200 °C temperature range changing its resistance value by a factor of 1000 in that interval. Devices are available commercially as very small thermistor beads below about 0.25 mm in diameter or as discs or rods. The beads have a very fast thermal response and are made by sintering a drop of thermistor paste onto two platinum alloy wires. The beads are then coated with glass to give a hermetic seal, or are mounted in an evacuated or glass filled tube.

The relationship between resistance and temperature of an NTC thermistor is given by

$$R_T = R_\infty \exp(B/T) \qquad (3.4)$$

where $R_T$ is the resistance at temperature $T$ K
$R_\infty$ is the resistance at an infinitely high temperature
$B$ is a constant of the device.

Equation (3.4) can be rewritten as follows

$$R_2 = R_1 \exp B \left( \frac{1}{T_2} - \frac{1}{T_1} \right) \qquad (3.5)$$

where $R_2$ is the resistance at $T_2$ K and $R_1$ is the resistance at $T_1$ K. The constant $B$ can have a value of up to about $10^4$ depending on the thermistor. This constant varies with temperature and can have as much as a 20% variation over the thermistor operating range, so that (3.4) and (3.5) are only useful over a limited temperature range. Manufacturer's data sheets usually quote the thermistor resistance at 298 K (25 °C) and the $B$ factor, so that the resistance at other temperatures can be found from (3.5).

Fig. 3.7 shows a plot of the typical variation of NTC thermistor resistance with temperature. (The PTC thermistor is described in section 3.5.2.) The resistance can change due to increase in the ambient temperature or due to the power dissipation in the thermistor itself. When the thermistor is used in applications where these

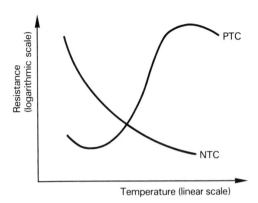

Fig. 3.7. Thermistor characteristics.

self-heating effects are not required the power dissipated by the thermistor must be kept low. Fig. 3.8 shows the voltage–current characteristics of a thermistor. At low values of current the curves are linear as the power dissipation is low and the resistance value remains constant for a given ambient temperature. The curves become non-linear as the self-heating effects become significant, until voltage $V_M$ is reached. After this point, the self-heating effects predominate and the resistance falls with increasing current, so reducing the voltage across the device. The resistance is now not appreciably affected by the ambient temperature.

Several parameters need to be considered when using NTC thermistors, and these are usually given in the manufacturer's data sheets. The power, usually quoted in milliwatts, needed

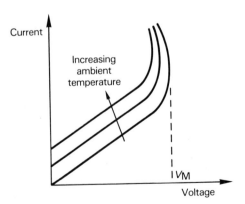

Fig. 3.8. Negative temperature coefficient (NTC) thermistor voltage–current characteristic.

to raise the thermistor's temperature by 1 °C is called the dissipation factor. The time taken to reach 63% of the total change in resistance in response to a step change in temperature is called the thermistor time constant. Thermistor sensitivity is usually quoted as a percentage change in resistance for a temperature change of 1 °C at a given ambient temperature. Thermistor drift is usually stated as a change in temperature for a given resistance value. A change of ± 0.5 °C after one year at 300 °C is typical, although if operated at below 150 °C this drift can be as low as ± 0.05 °C per year. There are two main reasons for drift, instability between the leads and the semiconductor material, and change in the impurity level and/or lattice structure of the semiconductor material. Causes of lead-semiconductor instability are: (i) different temperature coefficients of expansion between semiconductor and lead resulting in thermal stress; (ii) chemical reaction at the lead-semiconductor interface; and (iii) migration of the metal used for the contact. Thermistors which have sintered-in leads, such as beads, have a much better lead-semiconductor stability than those with metallised surface contacts such as discs. Generally discs are not suitable for use above about 150 °C although glass coated beads can be used at temperatures up to 300 °C.

In oxide semiconductors, the conduction is primarily due to impurities, so that the stability is very dependent on the environment to which the material is exposed. Therefore, for good stability thermistors require a hermetic seal.

### 3.5.2 PTC thermistors

The characteristic of a positive temperature coefficient (PTC) thermistor is more complex than the NTC thermistor and is shown in fig. 3.7. The resistance decreases initially until the Curie temperature of the PTC material is reached. There is then a sharp increase in resistance until a second negative temperature coefficient region is reached at a much higher resistance value. Because the PTC thermistor has a complex resistance-temperature relationship its operation cannot be defined by a simple formula, and data sheets often specify several points on the resistance-temperature characteristic. Also quoted on the data sheet is the switch tem-

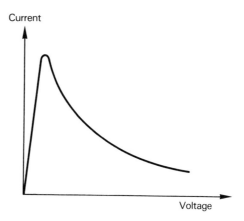

Fig. 3.9. Positive temperature coefficient (PTC) thermistor voltage–current characteristic.

perature, which is the temperature at which the thermistor resistance reaches twice its minimum value. The thermistor can be designed to give switch temperatures between about 20 °C and 200 °C.

PTC thermistors are usually made from barium titanate. In its monocrystalline form this material has a negative temperature coefficient. The PTC thermistor is manufactured from many such small crystals which are bonded together and sintered causing barrier layers to form at the intercrystalline boundaries causing resistance to the current flow. As the temperature increases so also does the resistance of these barriers, giving a material with a positive temperature coefficient.

Fig. 3.9 shows the voltage–current characteristic of a PTC thermistor. At low current levels the power dissipation is low and the curve is linear, the resistance being determined primarily by the ambient temperature. At a certain value of current the power dissipation within the thermistor is sufficient to raise its temperature into the switch mode. The resistance increases rapidly so that the current falls as the voltage increases further.

### 3.5.3 Voltage dependent resistors

The resistance of a voltage dependent resistor (VDR) decreases with increasing voltage which results in a device having a characteristic very similar to a back-to-back zener diode, as shown in fig. 3.10. In the low current leakage region

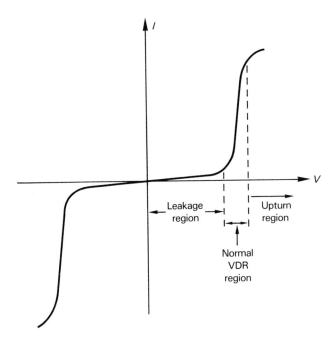

Fig. 3.10. Voltage dependent resistor (VDR) characteristic.

the curve is linear indicating a constant high resistance. This resistance decreases with temperature. At a certain voltage level the VDR enters its normal operating region. The resistance falls sharply causing a rise in current. This region follows the law

$$I = K V^\alpha \qquad (3.6)$$

where $I$ and $V$ are the current and voltage, and $K$ and $\alpha$ are constants. The values of these constants depend on the material and process used, and on the dimensions of the VDR. $\alpha$ has a value of about 25 to 100 and represents the figure of merit for the VDR. It is a measure of the non-linearity between two current points $I_1$ and $I_2$ and is given by

$$\alpha = \frac{\log(I_2/I_1)}{\log(V_2/V_1)} \qquad (3.7)$$

In the upturn region, the VDR resistance is relatively constant. The effect of temperature on the VDR characteristic is small beyond the leakage region.

Many materials have been used for VDRs, the most common being metal oxides. Because of this these devices are often called metal oxide varistors. The most commonly used material is zinc oxide. In the production of the VDR the zinc oxide is mixed with bismuth oxide and other metal oxides and pressed into shape and sintered. This results in a structure made up of minute grains, each separated from its neighbour by intergranular electrical barriers. These boundaries act as low voltage breakdown junctions whose voltage drop varies with the material used. For a zinc oxide based VDR the voltage drop is approximately two volts. The electrical properties of the VDR depend on the material composition and the dimensions. Capacitance effects also exist at the intergranular boundaries. The value of the capacitance between the electrodes of the VDR is inversely proportional to the number of granular boundaries involved in the material.

Several parameters are of interest when using a VDR, some of which are illustrated in fig.3.11. $V_x$ and $I_x$ are the voltage and current at any point which is specified on the data sheets. $V_{d.c.}$ and $I_{d.c.}$ are the maximum d.c. voltage and current which may be applied across the device, and $V_{a.c.}$, $I_{a.c.}$ and $V_p$, $I_p$ are the maximum a.c. values at 50 to 60 Hz and the

3.5   Non-linear resistors

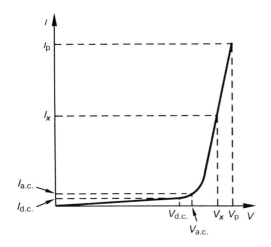

Fig. 3.11. Part of a VDR characteristic illustrating terms used.

maximum pulsed values respectively. The static resistance at any point is the ratio of the current and the voltage and is given by

$$R_s = V_x/I_x \qquad (3.8)$$

The dynamic impedance is the slope of the characteristic at that point and is given by

$$Z_D = d(V_x)/d(I_x)$$

which from (3.6) and (3.8) is equivalent to

$$Z_D = R_s/\alpha \qquad (3.9)$$

The voltage ratio $V_p/V_x$ is a measure of the VDR large signal non-linearity. Alternating current idle power is the power dissipation at $V_{a.c.}$. The leakage current is usually specified as $I_{d.c.}$. Maximum transient peak current is the maximum non-recurrent peak current from a random transient source. As the pulse width of the transient source increases, the peak allowable value of this current decreases. The maximum transient energy is the maximum power dissipation from a random transient pulse, and the maximum transient average power dissipation is the maximum average power dissipation due to a group of transient pulses.

Lead inductance of a VDR often results in resonance with the internal capacitance of the device. Even with no leads, a disc VDR will have a small self-inductance which gives a resonance frequency of about 100 MHz. The speed of response of a VDR is very fast, of the order of 1 ns since no heating effects are involved, as in a thermistor. Because of this short time the lead inductance and device capacitance effects are important and must be minimised. Generally when a rapidly rising voltage is applied across a VDR the voltage will overshoot the value at which it would have been clamped for a steady state voltage by between 5% and 25%. These overshoot effects are damped by the line impedance and VDR capacitance.

# 4. Capacitors

## 4.1 Introduction

Capacitors are the second most commonly used electronic components after resistors. In this chapter the terms used in specifying the characteristics of capacitors are introduced and followed by a description of the different types of commercially available capacitors.

## 4.2 Capacitor principles

### 4.2.1 *The equivalent circuit*

The basic construction of all capacitors is similar. They consist of two metal plates separated by a thickness of dielectric material. The capacitance ($C$) of this capacitor is given by

$$C = \epsilon_o \epsilon \, A/d \qquad (4.1)$$

where

$\epsilon_o$ is the permittivity of free space and has a value of $8.85 \times 10^{-11}$ F/m
$\epsilon$ is the permittivity of the dielectric
$A$ is the area of the plates
$d$ is the distance between the plates (i.e. the thickness of the dielectric).

For a large value of $C$ the capacitor must either use a thin dielectric with a high permittivity or have a large plate area. Problems can arise from the thin dielectric, high permittivity capacitor for two reasons. First, a thin dielectric is susceptible to stresses imposed on it during its lifetime, and second, materials with high permittivity tend to have characteristics which are voltage dependent.

The equivalent circuit of a capacitor is shown in fig. 4.1. Resistance $R_P$ is the leakage or insulation resistance of the capacitor and is described later in this section. Resistance $R_S$ is the series resistance. The a.c. resistance of the capacitor, which reflects the values of both $R_P$ and $R_S$, is usually referred to as the equivalent series resistance or ESR and represents the losses in the capacitor. The losses are of two types,

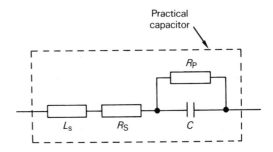

Fig. 4.1. Equivalent circuit of a capacitor.

the first is the true dielectric loss component and the second is attributable to the resistive components of the electrodes, leads and such like. $L_s$ is the self-inductance of the capacitor and leads, and it can generally be ignored at low frequencies.

Based on its low frequency equivalent circuit (i.e. $L_s$ is ignored), the following relationships can be obtained:

$$Z = (R_S^2 + X_C^2)^{1/2} \qquad (4.2)$$

$$\cos \phi = R_S/Z \qquad (4.3)$$

$$\tan \delta = R_S/X_C \qquad (4.4)$$

where

$Z$ is the impedance of the capacitor
$X_C = 1/2\pi f C$ and is the capacitor reactance at frequency $f$
$\cos \phi$ is called the power factor of the capacitor
$\tan \delta$ is the tangent of the loss angle, also known as the dissipation factor of the capacitor.

The power factor can be used to find the proportion of the a.c. signal which is dissipated as heat in the body of the capacitor. For low values of series resistance $R_S$ the angle $\delta$ is small and the power factor is approximately equal to the dissipation factor, this being the case for most practical capacitors. Power factor and dissipation factor vary in a complex manner with environmental conditions. For example, the power factor decreases with decreasing temperature and falls sharply at temperatures below 0 °C; the dissipation factor is sensitive to the presence of moisture.

### 4.2.2 *Capacitor ratings*

The absolute maximum ratings of a capacitor are specified in terms of voltage, current and power dissipation.

The d.c. voltage rating is determined primarily by the maximum voltage gradient which the dielectric can stand. The d.c. voltage rating is also called the insulation breakdown voltage. In some types of capacitor constructions the electrodes extend beyond the dielectric and there is the possibility of a voltage breakdown over the surface of the dielectric rather than through it.

The a.c. characteristics of a capacitor depend on frequency. At low frequencies the peak of the a.c. voltage should be kept below the maximum d.c. voltage rating. At higher frequencies hysteresis losses occur in the dielectric which cause a temperature rise and this effect may set limits on the root mean square (rms) voltage rating of the capacitor in a given frequency and ambient temperature range. The surge voltage rating is often greater than the d.c. rating provided the surge is of short duration. Because at higher ambient temperatures the permissible internal power dissipation must be kept lower the data sheets usually also specify a voltage derating factor (units: volts per degree centigrade).

The current rating of the capacitor is usually specified as the rms current and is limited by temperature rises in the capacitor. Current loss occurs due to hysteresis effects in the dielectric. In the equivalent circuit (fig. 4.1) this effect is represented by the parallel resistor $R_p$. The maximum current rating is the value of the current which would cause the dielectric to reach its maximum temperature. If this temperature is exceeded the capacitor could be damaged. The capacitor can be specially designed to dissipate its internally generated heat and so increase its current rating. The surge current rating is also usually specified on data sheets. For capacitors used as reservoirs of charge, such as in power supply applications, the rating of the current which flows in and out of the capacitor, called the a.c. ripple current rating is an important parameter as the ripple current causes heating and can lead to failures or shorten the capacitor life. This current is normally quoted in capacitor data sheets at a specific temperature and frequency, but in most circuit applications it is very difficult to predict accurately. It is usual in these instances to make an initial estimate and then to measure the ripple current, when the capacitor is in circuit, using a true rms meter.

### 4.2.3 *Capacitor characteristics*

Many capacitor characteristics are dependent on the properties of its dielectric. The dielectric strength is the maximum voltage gradient that can be applied to the dielectric before breakdown. It is measured in volts per metre and depends on the particular dielectric material used and its thickness. The dielectric strength is related to the dielectric or insulation breakdown voltage since the greater the applied voltage the higher the electric field within the dielectric material and the greater the probability of breakdown. The product of dielectric strength and dielectric thickness gives the maximum working voltage of the capacitor. The dielectric strength is affected by factors such as moisture, high temperature, frequency and the wave shape of the applied voltage.

In an ideal capacitor no energy should be dissipated in the dielectric as a result of the applied voltage. In all practical capacitors there is dissipation and this is called dielectric loss and is considered to be caused by an insulation resistance. When the applied voltage is d.c. the resulting loss is sometimes attributed to the leakage current and the resistance is referred to as the capacitor leakage resistance. The insulation resistance of the capacitor is determined by factors such as the volume resistivity of the dielectric, the capacitor encapsulation, surface contamination, environmental conditions and so on. The higher the insulation resistance the longer a capacitor can store a charge. Insulation resistance decreases with temperature and with some types of capacitors the insulation resistance is so low that it is more common not to quote it at all in manufacturers' data sheets, but to refer instead to the leakage current.

Another dielectric characteristic is called the dielectric absorption. When a fully charged capacitor is discharged and allowed to remain open circuit for a period of time it will accumulate a new charge which is equal to a fraction of the original charge. This is the charge absorbed by the dielectric and it causes a time lag between the rate of charging and discharging of the capacitor. At high frequency operation the lag in discharging means that the capacitor cannot complete its discharge, and this has the same effect as a loss in the value of capacitance. The dielectric absorption is usually stated as a ratio of the voltage to which a capacitor recovers after a discharge to the original charging voltage, expressed as a percentage. The energy stored in a capacitor $C$ charged to a d.c. voltage $V$ is given by

$$E = \tfrac{1}{2} CV^2 \qquad (4.5)$$

Since a hysteresis loop effect exists between the charge and the applied voltage, and this results in energy dissipation per cycle, the total energy stored in a capacitor when the voltage is a.c. is less than that given by (4.5).

Substituting for $C$ from (4.1) into (4.5) gives

$$E = \tfrac{1}{2} \epsilon_o \{ \epsilon (V/d)^2 \} Ad \qquad (4.6)$$

The first term, $\tfrac{1}{2}\epsilon_o$, in this expression is a constant, the second term, $\{\epsilon(V/d)^2\}$, is dependent on the properties of the capacitor dielectric and the third term, $Ad$, represents the volume of the capacitor. Therefore in order to store the maximum energy in a given volume the dielectric should be chosen such that the second term in (4.6) is maximised.

Capacitors are selected after construction according to their tolerances about the nominal value, as is done for resistors. However, whereas a 5% tolerance is fairly easily attainable for most resistors, some capacitors have selection tolerances of as much as 100%, and during use the capacitor may drift even further from the selected value. The long term stability of the capacitor is a measure of the change in capacitance during its operating life. It is specified as the percentage change after a given number of hours of operation at a given temperature. The temperature coefficient of capacitance quantifies change in capacitance with temperature and is specified on data sheets in parts per million change per degree centigrade (ppm/°C) or in percentage change per degree centigrade (%/°C).

The temperature coefficient of a capacitor is dependent primarily on the type of dielectric used but it can also be affected by the way the capacitor is constructed. For example, a difference in expansion between the encapsulant and element can give a change in capacitance. Some dielectrics also have different temperature coefficients over different temperature ranges. The temperature coefficient of capacitance may be positive or negative in the range 60 ppm/°C to 6000 ppm/°C.

Noise is not as important a parameter in a capacitor as in a resistor, and it is not always specified on data sheets. The noise is usually caused by variable contact resistance in the connections of the terminals to the capacitor electrodes, and by poor metallisation of the dielectric which results in isolated islands of electrodes or parts which are only partially connected to the terminals.

Capacitor characteristics are affected by the operating frequency. At very low frequencies the d.c. leakage and time constant effects, and the leakage through the encapsulation, are important. As the frequency increases the parallel loss resistivity ($R_p$ in fig. 4.1) becomes important and at very high frequency the impedance of the dielectric ($C$ in fig. 4.1)

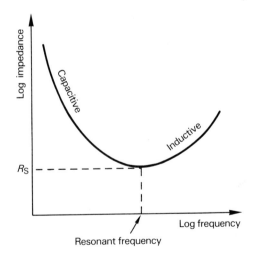

Fig. 4.2. Frequency characteristic of capacitor impedance.

shunts $R_P$ and the series effects of resistance ($R_S$) and inductance $L_s$ begin to dominate. Fig. 4.2 illustrates the capacitor frequency/impedance curve based on the equivalent circuit of Fig. 4.1. The resonant frequency is a much quoted data sheet parameter and can vary from 10 kHz to 10 MHz depending on the type of capacitor.

## 4.3 Electrolytic capacitors

Electrolytic capacitors have the highest capacitance–voltage ($CV$) product in a given case size as well as the largest absolute capacitance value. The large capacitance values are attainable because of the use of a very thin dielectric film formed by oxidising a metal (usually aluminium or tantalum). There are two types of electrolytic capacitors, based on aluminium and tantalum, and these are described in sections 4.3.1 to 4.3.3.

### 4.3.1 *Construction of aluminium capacitors*

One electrode of the capacitor, usually called the anode, is made from a thin foil of aluminium. This is oxidised to give a thin coating of aluminium oxide which acts as the dielectric. The other electrode of the capacitor is formed by an electrolyte, usually glycol borate and is called the cathode. For low equivalent series resistance (ESR) capacitors non-aqueous electrolytes are used while to improve ripple characteristics and temperature range ethylene glycol is preferred. The cathode is in contact with a second aluminium foil called the cathode plate and the connections to the outside of the capacitor are made from this plate. In a practical device the anode and cathode plates are separated by a layer of very porous paper which is impregnated with the electrolyte. The foils and paper are wound together and placed in a metal can as shown in fig. 4.3 Since the electrolyte makes contact with the housing it is at cathode potential so that if the can needs to be insulated it is often protected by sleeving. It is also common practice to connect the wound foil assembly to the case to increase the heat conduction. The inductance of the capacitor depends largely on the lead connections, and to achieve low inductances multiple connections are made to several parts of the foil along its length, and these are then joined together.

The production of the aluminium electrolytic

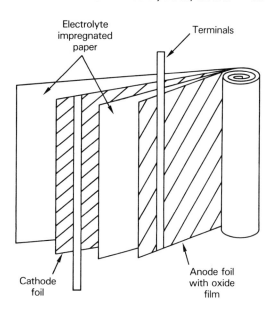

Fig. 4.3. Aluminium electrolytic capacitor construction.

capacitor consists of several steps. During the preparation stage the anode foil is inspected to ensure that it reaches a purity of at least 99.99% aluminium. Impurities such as iron, copper, manganese, boron and silicon will prevent good oxide formation whilst other impurities such as sulphate or chloride ions will effect the stability of the capacitor. The aluminium foil may be brushed or chemically cleaned at this stage to form nucleation sites for any subsequent foil etching. Brushing also makes the foil less brittle so that it can be wound more easily.

The anode foil may now be etched to give an etched foil structure. Differences between etched and plain capacitors are described later. The anode foil is then immersed in a tank of electrolyte, the cathode in the tank usually being formed by an aluminium or steel plate. When a positive voltage is applied to the anode an oxide growth of about 1.5μm thickness forms on it, the thickness of the oxide being dependent on the magnitude of the applied voltage. This process is known as anodisation.

After anodisation the anode foil is cut to the correct width, the terminals attached and it is wound with cathode foils and dry paper. The paper is then impregnated, usually in vacuum, with the electrolyte. The choice of electrolyte depends on the intended working voltage of the capacitor and can vary in resistivity from 200 $\Omega$ cm to 2000 $\Omega$ cm. The capacitor is now 'aged' by applying a voltage in excess of its intended working voltage. This voltage also repairs the damage caused during cutting and terminal attachment. Finally the assembly is encapsulated in a can.

In order to increase the capacitance value attainable from a given physical size the aluminium foil may be etched. This increases the area of the dielectric and also increases the relative permittivity from about 7 to 10. The etched surfaces must not be filled in subsequently by the oxide, therefore there is a limiting thickness of oxide that can be grown and still preserve the surface gain obtained by etching. This limits the maximum operating voltage of this type of capacitor.

The electrolyte is the true negative electrode or cathode of the electrolytic capacitor. Since it is an electrolyte the cathode fits well to the etched anode surface. The electrolyte serves another vital purpose in healing any weakness which may arise in the oxide film during operation. The dielectric is slightly attacked by the electrolyte but the lost oxide is regrown when the capacitor is in use. This regrowth releases gases which are generally absorbed back into the electrolyte. A safety vent is usually fitted to the capacitor can to allow gas to escape during short term overloads.

In an electrolytic capacitor the cathode foil is usually highly etched and only lightly anodised, if at all. This gives the cathode a very high capacitance compared to the anode and since the two foils are in series the effect of the cathode foil on the total capacitor can usually be ignored. Fig. 4.4 shows the equivalent circuit of an aluminium electrolytic capacitor. The most important elements in this are the anode dielectric capacitance and the electrolyte resistance. The effect of the series resistance elements of the capacitor is to increase the operating losses and so limit the maximum a.c. current rating. Losses due to the electrolyte are higher in an etched foil capacitor than in a plain foil capacitor because of the increase in series resistance caused by the etched and anodised grooves on the surface of the foil.

### 4.3.2 Characteristics of aluminium capacitors

For an electrolytic capacitor the equivalent series resistance (ESR) is often quoted instead of the dissipation factor, tan $\delta$. The ESR is made up of the resistance of the electrodes, anode and cathode foils, connections to the foils and the resistance due to dielectric loss. As the operating temperature decreases the electrolyte resistance increases but the foil and dielectric losses decrease. For low voltage etched foil capacitors the electrolyte resistance dominates so that the ESR increases with decreasing temperature. For large capacitors there is only a small change in ESR with temperature since the electrolyte resistance does not predominate. The electrolyte resistance is also relatively independent of frequency so that the effect of frequency on ESR for low voltage etched foil capacitors is small. For high voltage or smooth foil capacitors the ESR increases with frequency due to the effect of frequency on the dielectric loss. The impedance–frequency curve is similar to that shown in fig. 4.2. The dissipation factor of the capacitor increases with frequency and decreases with temperature, and the capacitance increases with temperature. At high temperatures the electrolyte resistance

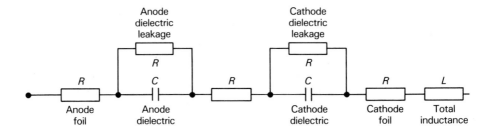

Fig. 4.4. Equivalent circuit of an aluminium electrolytic capacitor.

decreases and this results in an increase in capacitor leakage.

If an electrolytic capacitor is stored or operated at low voltages for long periods it tends to form to the lower voltage, and if the capacitor voltage is then raised to the rated value large currents will flow and gassing will occur. To reform the capacitor back to its rated value the rated voltage may be applied to it through a series resistance until the current falls to a low value. Temperature has a marked effect on capacitor storage life. If the storage temperature is below 40 °C then the electrolytic capacitor can usually be stored for about three years without the need to reform it before use.

The life of an electrolytic capacitor is very dependent on the type of electrolyte used, and the end of life is marked by a decrease in capacitance and an increase in impedance and tan $\delta$. Loss of capacitor life is also caused by exhaustion of the electrolyte. This arises from the gradual conversion of electrolyte into gas and the diffusion of the gas through the capacitor seal during operation. For long life capacitors, the electrolyte content must be high, the capacitor must be well sealed to maintain a vapour pressure over the liquid to prevent drying out, and the capacitor should be run at low temperatures.

Electrolytic capacitors are usually polarised, which means that they must always be operated with the anode positive relative to the cathode. Applying a reverse voltage causes an electrolytic process to occur which generates an oxide film on the cathode giving excessive heating and gas formation which can damage the capacitor. The increase of oxide film on the cathode also reduces capacitance, and since it is in series with the anode foil the overall capacitance value is reduced. Generally, the cathode plate is covered by a very thin oxide layer due to atmospheric action which enables the capacitor to withstand about 2V reverse voltage without adverse effect. Non-polarised or bipolar electrolytic capacitors are available but these consist of two foils which are anodised during production such that the capacitance due to either electrode is the same. These foils are connected in series so a non-polar capacitor has about half the capacitance value of a polar capacitor of the same volume.

### 4.3.3 *Tantalum electrolytic capacitors*

These capacitors use tantalum oxide as the dielectric, which has a higher relative permittivity than aluminium oxide (25 compared to 10) and so can give a larger capacitance value in a given can size. The shelf life of tantalum capacitors is also better than that of aluminium capacitors since a purer dielectric can be formed. However, tantalum is more expensive and heavier than aluminium. Tantalum can be etched, but the capacitance gain obtained by this method is lower than that obtained with aluminium.

There are three basic types of tantalum capacitors, tantalum foil, solid tantalum and wet sintered tantalum. Tantalum foil capacitors are made by oxidising an anode tantalum foil to form the dielectric oxide, and then winding it with a second tantalum foil using a construction similar to that used in aluminium electrolytic capacitors. The electrolyte used is usually sulphuric acid.

The solid and wet sintered tantalum capacitors differ mainly in the type of cathode system used. A similar anode is used for both. The anode element is made of porous pellets obtained by compressing tantalum powder with a grain size of 4 $\mu$m to 10 $\mu$m in a vacuum at 1500 °C to 2000 °C. This causes an oxide growth, which forms the dielectric, on the individual grains. The capacitance value is determined by the sintering time and temperature, the smaller grain size having higher surface gains at lower temperatures. The high sintering temperatures are useful in removing any surface impurities. A higher sintering time–temperature product will produce lower leakage current and a decrease in the capacitance value. Tan $\delta$ increases with an increasing time–temperature product due to the fine narrow channels which are formed in the material as sintering increases.

For a solid tantalum capacitor, the porous anode pellet obtained after sintering is impregnated with manganese nitrate and heated to 400 °C which decomposes the manganese nitrate to form manganese dioxide. This oxide forms a solid electrolyte negative electrode and makes contact with 90% to 98% of the surface of the dielectric. Control of this contact area during production is very important since it affects both the batch-to-batch and within-a-batch capacitance values. The solid manganese dioxide electrolyte has several advantages over conventional liquid electrolytes; for example there is no drying out due to evaporation or leakage so that ageing effects are slight and operation over a broader frequency band is possible. This results from there being electron conduction in the solid electrolyte rather than ion conduction, as in a liquid electrolyte, which gives a lower specific resistance and better frequency performance. The manganese dioxide has a self-healing effect similar to that of the electrolytes used in aluminium capacitors.

The magnesium dioxide is coated with a layer of graphite and silver and encapsulated in the cathode system, which can be of many types. In the slug construction, shown in fig. 4.5a, the coating is dipped in epoxy resin whereas the structure shown in fig. 4.5b uses a metal case with a resin seal. (Solid tantalum capacitors have a tighter tolerance than aluminium) and can operate over a wider temperature range due to the greater stability of the tantalum pentoxide film.

The wet sintered tantalum capacitor uses an

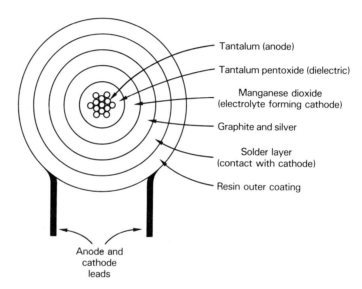

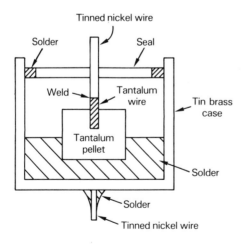

Fig. 4.5. Solid tantalum capacitor construction; (a) dipped resin coated, (b) metal can.

anode slug similar in construction to that of a solid tantalum capacitor. The electrolyte can be either a mixture of sulphuric acid and deionised water, or a gelled electrolyte which is usually a mixture of sulphuric acid and silica. Gelled electrolytes maintain the same high conductivity, low electrical losses and high ripple current as liquid electrolytes but give a lower leakage current. Fig. 4.6 shows the construction of a typical wet tantalum capacitor.

The cathode system used in a tantalum capacitor needs careful consideration since it usually makes up the case and therefore often determines parameters such as shelf life. Three types of systems are in use, silver cathode, platinised silver cathode and sintered tantalum cathode. The silver cathode is only occasionally used. It has the disadvantage that the silver can dissolve and collect in the electrolyte or deposit onto the anode so that the cathode surface is disrupted, the capacitance value reduced, and the leakage current increased which may ultimately lead to a short circuit. These disadvantages are overcome in platinised silver cathodes by treating the internal surface of the silver case with black platinising. This also increases the capacitance per unit area of cathode by a factor of ten to fifteen. The sintered tantalum cathode capacitor has the highest $CV$ product per volume. The cathode material is similar to that used for the anode and the capacitor can withstand about three volts in the reverse direction. In all types of tantalum capacitors, an over-voltage in the forward direction will cause the dielectric to grow in thickness reducing the capacitance value. In the limiting case dielectric breakdown will occur.

Fig. 4.7 illustrates the basic differences between the three types of tantalum capacitors. The operating temperature range of all types is usually -55 °C to +125 °C but in the wet sintered system it is possible to extend this to +200 °C by using special containers which permit the electrolyte to be used above the boiling point. The solid tantalum capacitor has a linear capacitance–temperature characteristic but both the other types demonstrate a rapid fall in the mobility of the electrolyte ions. The frequency performance of the capacitors is also dependent on electrolyte conduction and since the ion mobility in liquid systems is low the capacitor's high frequency performance is reduced.

Although wet tantalum capacitors have a high $CV$ product it is not practical to make them in small physical sizes due to the space needed by seals, etc. It is also very difficult to make solid tantalum capacitors in large capacitance values, and these devices are more suited to mass production in small values and sizes.

Solid tantalum capacitors are reliable and

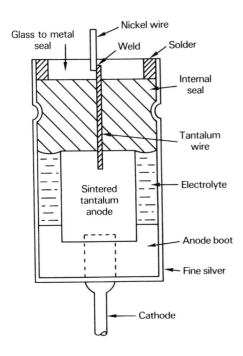

Fig. 4.6. Wet sintered tantalum capacitor.

| Parameter | Foil | Wet | Solid |
|---|---|---|---|
| Maximum d.c. rating | 450 V | 125 V | 100 V |
| Best available tolerance | ±10% | ±5% | ±5% |
| Temperature stability | 3 | 2 | 1 |
| Frequency characteristic | 2 | 2 | 1 |
| Cost | 1 | 2 | 3 |
| $CV$ per unit volume | 3 | 1 | 2 |
| Life | 3 | 2 | 1 |
| Leakage current | 2 | 1 | 3 |

Fig. 4.7. Comparison of different types of tantalum capacitors; 1 = best, 3 = worst.

have a long life because there is no wear-out mechanism such as the loss of electrolyte. Failure in these capacitors is primarily due to oxide crystallisation. Wet and foil tantalum capacitors fail due to evaporation through the seal causing a fall in capacitance value and a degradation of the dissipation factor.

## 4.4 Paper and plastic capacitors

Paper and plastic film capacitors are often referred to as wound capacitors since they are made by winding dielectric and conductive material together in a way similar to that shown in fig. 4.3. The most commonly used dielectric material used to be paper, but this has now been largely replaced by plastic film dielectrics, except for capacitors with very high d.c. voltage or low frequency a.c. power applications.

### 4.4.1 *Paper capacitors*

Paper has pores which contain moisture totalling up to about 8% of the weight of the paper. During the manufacturing process the water is evaporated by vacuum drying and the pores are impregnated under vacuum with a mineral, or a synthetic liquid, or a resin dielectric. There are still conducting particles left in the paper, however, and these are often minute holes in the material so that generally at least two layers of paper dielectric are used between each metal foil layer. The disadvantage of using liquid impregnated paper is that it needs to be assembled in sealed containers to prevent seepage. The advantage is that it gives the capacitor a higher ionisation voltage level. The power factor of the capacitor is dependent on the type of impregnant used but is generally large and increases rapidly with frequencies above 10 kHz.

Two types of construction are used for paper capacitors, foil and film and metallised film. The foil and film capacitor is made by interleaving alternate layers of metal foil and impregnated paper and winding these together. This type of construction is used for high voltage and high current applications. In the metallised film capacitor, the impregnated paper dielectric is coated with a thin layer of aluminium or zinc to form the metal plate and several layers are then wound. Zinc corrodes easily and has a lower vapour pressure so that modern capacitors almost exclusively use aluminium, the layer having a resistivity of about 1.5 to 3.0 $\Omega$ per square.

Metallised film capacitors have smaller physical sizes than foil and film types since the metal layer has a thickness of 0.01 $\mu$m to 0.1 $\mu$m compared to a foil thickness of 5 $\mu$m. The metallised film capacitors have the added advantage that they are self-healing. A defect in a dielectric, such as a minute hole, would result in an arc between electrodes at high voltage. In a foil type capacitor this arc would destroy the surrounding dielectric and result in a short circuit. In a metallised film capacitor the heat of the arc evaporates some of the thin metal layer surrounding the hole and so clears the short circuit. The energy stored in the capacitor must be sufficient to effect clean self-healing, especially when the circuit impedance is such that no more external energy is available, and the design and manufacturing process used for the capacitor achieve this. The main disadvantages of metallised film capacitors are that they have a low pulse rise time and current rating. A metal foil capacitor has a pulse rating of about 800 V/$\mu$s compared to 20 V/$\mu$s for metallised types. The current rating of a capacitor depends on the current carrying capability of its conducting parts and on its internal temperature rise, which is determined by the heat dissipation ability of the construction. In both these factors the foil capacitor is more efficient than the metallised version and hence has a higher current rating for the same size.

### 4.4.2 *Plastic film capacitors*

The most commonly used types of plastic materials are shown in fig. 4.8. The permittivity is frequency dependent such that the capacitance value decreases by about 3% for

| Film | Relative permittivity at 1 kHz | Film thickness ($\mu$m) |
|---|---|---|
| Polystyrene | 2.4 | 8 |
| Polyester | 3.3 | 3.5 |
| Polycarbonate | 2.8 | 1.5 |
| Polypropylene | 2.3 | 8 |
| Paper | 2 to 5 | 5 |

Fig. 4.8. Characteristics of plastic films.

| Parameters | Electrolytic Aluminium | Electrolytic Tantalum (solid and wet) | Paper Foil | Paper Metallised | Polyester Foil | Polyester Metallised | Polycarbonate Foil | Polycarbonate Metallised | Polystyrene | Polypropylene Foil | Polypropylene Metallised | Ceramic Disc and tubular | Ceramic Multilayer | Mica |
|---|---|---|---|---|---|---|---|---|---|---|---|---|---|---|
| Capacitance range ($\mu$F) | 0.5 to 100 000 | 0.1 to 1000 | 0.001 to 100 | 0.001 to 200 | 100 pF to 0.01 $\mu$F | 0.001 to 10 | 5 pF to 0.01 $\mu$F | 0.001 to 100 | 100 pF to 1 $\mu$F | 100 pF to 1.0 $\mu$F | 0.001 to 100 | 5 pF to 1 $\mu$F | 10 pF to 10 $\mu$F | 1 pF to 1 $\mu$F |
| Voltage (V) d.c. / a.c. | 6.3 to 500 | 1 to 100 | 100 to 5000 / 250 to 1000 | 100 to 5000 / 250 to 1000 | 150 to 400 / 100 to 160 | 100 to 1500 / 63 to 400 | 100 to 500 / 63 to 160 | 63 to 1000 / 40 to 250 | 63 to 1000 | 100 to 1500 / 63 to 500 | 750 to 1000 / 250 to 500 | 63 to 1000 / 63 to 250 | 63 to 500 | 63 to 630 |
| Tolerance (%) | 20 | 5 | 5 | 10 | 5 | 5 | 2 | 5 | 0.5 | 2 | 5 | 10 | 10 | 0.5 |
| Tan $\delta$ | 0.08 | 0.005 to 0.02 | 0.005 | 0.01 | 0.005 | 0.01 | 0.001 | 0.005 | 0.0003 | 0.001 | 0.001 | 0.002 | 0.02 | 0.005 |
| Insulation resistance (M$\Omega$) | Very low | Very low | $2 \times 10^4$ | $3 \times 10^3$ | $10^5$ | $5 \times 10^4$ | $10^5$ | $5 \times 10^4$ | $10^6$ | $5 \times 10^4$ | $10^5$ | $10^2$ | $10^4$ | $10^5$ |
| Temperature coefficient (ppm/°C) | 1500 | 100 to 1000 | 300 | 300 | 400 | 400 | 150 | $-100$ | $-150$ | $-100$ | $-200$ | Non-linear positive to 2000 negative | | 100 |
| Stability (comparative) | 3 | 2 | 4 | 4 | 4 | 4 | 4 | 4 | 1 | 2 | 4 | 4 | 4 | 1 |
| Relative size per CV product (comparative) | 1 | 2 | 3 | 2 | 2 | 2 | 2 | 2 | 3 | 2 | 2 | 2 | 2 | 2 |
| Approximate resonance (MHz) | | | 0.1 | 0.1 | 1 | 0.1 | 1 | 0.1 | 1 | 1 | 0.1 | 10 | 100 | 1 |
| Relative cost per CV product (comparative) | 2 | 3 | 2 | 2 | 2 | 1 | 2 | 2 | 3 | 1 | 1 | 1 | 1 | 2 |

Fig. 4.9. Summary of capacitor characteristics. In the comparative data 1 = best or lowest, 4 = worst or largest.

each decade increase in frequency. Both foil and film and metallised constructions are used and the capacitors are wound as shown in fig.4.3, or they are wound flat to give a flat rectangular shape which allows higher packing densities on a printed circuit board. Metallised plastic capacitors have almost replaced paper capacitors for low voltage applications since they are physically smaller and have better electrical characteristics. Plastic film cannot be impregnated efficiently because the synthetic film is not sufficiently wetted by the impregnant. Plastic capacitors are therefore of dry construction which limits their peak a.c. and d.c. voltage rating.

Polystyrene was one of the earliest plastic films to be used. Polystyrene capacitors are still widely used in applications requiring low loss, high insulation resistance, close tolerance and good stability. These parameters are obtained due to the stable electrical characteristics of the polystyrene material and the ability to wind the capacitor to close tolerances. Metallised polystyrene capacitors are not commonly available since polystyrene has a low melting point making it difficult to vacuum deposit a metal film onto it. Polystyrene has low water absorption up to about 80% relative humidity. The material contains air spaces, but most of these are removed by heat processing during manufacture which also results in controlled shrinkage so that the capacitor is rigid and robust enough to be used in an unencapsulated state. However, the material is affected by high temperature, grease and solvents, such as encountered during printed circuit board flow soldering, so it is usually encapsulated.

Polyester films are usually of the polyethylene type, and are obtained under various trade names such as Mylar (Du Pont) and Melinex (ICI). They have good electrical characteristics, high permittivity, and low film thickness, and they are relatively cheap so that they are widely used for general purpose d.c. applications.

Metallised polycarbonate films are used in applications where size is important. Their electrical characteristics are similar to but slightly better than polyester.

Polypropylene is a relatively new capacitor material. It is cheap and has similar properties to polystyrene but it can be used at higher temperatures so that it can be metallised. Its low cost per unit of $CV$ product makes it especially attractive for a.c. mains applications such as fluorescent lighting.

There are many different encapsulants available for plastic film capacitors. These include dipped, compression or cast moulding, plastic sleeves or cases with resin end seals. Since the plastic films are less hygroscopic than paper they need less protection than given to paper capacitors.

Fig. 4.9 summarises the main parameters of paper and plastic film capacitors. Many of these parameters are frequency and temperature dependent. Fig. 4.10$a$ shows the capacitance change with temperature and polystyrene is seen to be the most stable. The dissipation factor of the capacitors is mainly dependent on the dielectric loss of the plastic foil and the resistance of the leads, contacts and foils. The lead, contact and foil resistance is usually kept small by using heavier metallisation at the edges of the layers. The dissipation factor increases with frequency and capacitance value, and fig. 4.10$b$ shows its variation with temperature. Polystyrene has the lowest dissipation factor and it is relatively constant. Insulation resistance, which may be specified as an absolute resistance value or as a time constant in seconds, equal to megaohms times microfarads, is the resistance of the dielectric between the layers plus the resistance between layer and case. It is determined by the quality of the materials used in the dielectric, case, mouldings and lead-throughs, as well as by the length of the surface leakage paths. Polystyrene has the highest insulation resistance as shown in fig. 4.10$c$ but this falls rapidly at high temperatures, indicating the unsuitability of this type of capacitor at these temperatures.

## 4.5 Ceramic capacitors

The dielectric for these capacitors is derived from the ferroelectric group of ceramic materials, mainly barium titanate ($BaO-TiO_2$). The actual composition of the dielectric can be varied widely to obtain a variety of different capacitor characteristics. Barium titanate has a high degree of polarisation which gives it a large dielectric constant, but this also makes it sensitive to frequency and voltage variations.

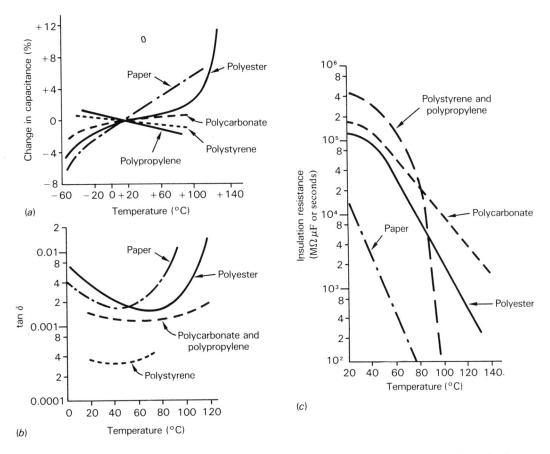

Fig. 4.10. Paper and film capacitor characteristics; (a) capacitor change with temperature, (b) dissipation factor (tan δ) change with temperature, (c) insulation resistance change with temperature.

### 4.5.1 Capacitor manufacture

The first step in the manufacture of the ceramic capacitor is the preparation of a mixture of fine ceramic powders and resin binders. This forms a flexible material which is known as green ceramic and it is cast into sheets or cylinders of the required thickness. The conductor electrodes are then silk screened or painted onto the green ceramic. The conductors are generally made from precious metal paste such as platinum, palladium, or silver combinations since these have high melting points and low oxidisation properties and can withstand the high temperatures which are subsequently used. The individual green plates may now be cut into the desired capacitor shapes such as squares and discs, or they may be first stacked together and pressed, as is done in the construction of multilayer capacitors, and then cut into shapes. The multilayer structure is also referred to as a monolithic capacitor and in it the successive electrodes extend from alternate sides, so that the individual layers are connected in parallel.

The cut green ceramic dice, i.e. the cut shapes, are now baked to remove the organic binders, and then sintered at a temperature of between 1200 °C and 1500 °C. This fuses the individual layers of a multilayer capacitor together. Care must be exercised during the sintering stage since poor firing will result in microcracks and delamination of the dielectric, and will give a capacitor with a low capacitance value, electric strength and insulation resistance.

The ends of the ceramic die are next dipped into a precious metal paste and baked at about

800 °C to form solderable terminations. Leads are connected to these terminations in axial or radial form. Several packaging methods are then used. The die can be moulded using transfer moulding or it can be potted by putting it into a pre-moulded epoxy case which is then filled with liquid resin. Glass encapsulations are also used in which the die is sealed in a glass tube. In the dual-in-line package construction the die is soldered to the lead frame and the assembly is then potted.

Conformal coating is widely used for encapsulating ceramic dice. A heated die may be dipped into the liquid resin or into a fluidised powdered resin bed where air under pressure is blown through the powder. An alternative technique uses a high potential electrostatic field to attract powder particles onto the die instead of dipping it into the powder. The die does not now need to be heated. In both techniques the die is cured in an oven after it has been covered with the resin powder. Conformal coating gives a cheap encapsulation method but it does not enable close tolerances to be maintained on the capacitor dimensions. The technique is, however, efficient in excluding moisture from the die, which is necessary since moisture affects the capacitance value, dissipation factor, insulation resistance and breakdown voltage of the capacitor.

**4.5.2** *Capacitor characteristics*

Ceramic capacitors are usually divided into two classes as shown in fig. 4.11. Class 1 capacitors use a very low proportion of barium titanate so that they have a low dielectric constant and capacitance per unit volume, but they have relatively stable characteristics. Class 2 capacitors use a greater proportion of barium titanate and high permittivity materials. Class 2a types are semi-stable and are used for high reliability applications whereas class 2b (also called high $K$) types use material with high dielectric constant, and are used for decoupling applications where large capacitance per unit volume is needed but the characteristics need not be very stable.

Ceramic materials are affected by voltage, temperature and frequency, the extent of this being determined by the material used and the construction. Capacitance change with temperature is non-linear. The temperature coefficient given in data sheets is obtained by dividing the total variation in capacitance by the temperature change. The temperature coefficient can be positive or negative and usually the higher the dielectric constant the more negative the temperature coefficient.

The effect of d.c. voltage is to cause a disorientation of the cells within the polarisation domains of the ceramic material which results in a reduction of capacitance value. The thinner the dielectric material the greater the electrical stress and the greater the effect of the applied voltage, so for high capacity multilayer ceramics the effect of d.c. voltage is greater than for high voltage single layer capacitors. Class 2 capacitors have a much larger voltage coefficient than class 1 devices. An applied a.c. voltage causes a reorientation in the polarisation domains within the dielectric and has the opposite effect to a d.c. voltage so that the capacitance value increases.

The capacitance of ceramic capacitors decreases with increasing frequency and the extent of this change depends on the capacitance

| Parameter | Class 1 | Class 2 a | Class 2 b (high $K$) |
|---|---|---|---|
| Dielectric constant at 25 °C | 5 to 500 | 200 to 3000 | 3000 to 10 000 |
| Operating temperature range (°C) | −55 to +125 | −55 to +125 | −55 to +85 |
| Temperature coefficient (ppm/°C) | ±30 | ±100 | −500 to −10 000 |
| Voltage coefficient at 25 °C | Negligible | −20% | −80% |
| Dissipation factor at 25 °C and 1 kHz | 0.001 | 0.025 | 0.03 |
| Ageing per decade (%) | Negligible | 0.5 to 3.0 | 2 to 10 |

Fig. 4.11. Characteristics of class 1 and class 2 ceramic materials.

value, body and lead dimensions and materials used. There is negligible change in capacitance for class 1 devices but a large change for class 2 capacitors.

The dissipation factor of ceramic capacitors is often used during the production stage to check on the quality of the processes, e.g. the consistency of the ceramic formulation, contamination, defective construction and so on. For class 1 capacitors the dissipation factor is stable, but for class 2 it decreases with temperature and d.c. voltage and increases with frequency and applied a.c. voltage. The insulation resistance of a ceramic capacitor is very dependent on the composition of the ceramic material. Usually, the lower the dielectric constant the higher the insulation resistance. The resistance decreases as the temperature increases for all classes of capacitors.

Ceramic capacitors are subject to ageing, which results in a decrease of capacitance and dissipation factor with time. This ageing is due to a gradual re-alignment of the crystalline structure of the ceramic and it is accelerated when the material is subjected to stress such as by operating near its rated d.c. voltage. The losses can be recovered by heating the capacitor above its curing temperature and this is known as de-ageing. Since ageing occurs in a logarithmic fashion with time very little change takes place after 1000 hours. Therefore capacitors can be pre-aged prior to use by applying d.c. voltage for a specific time. Class 1 capacitors show negligible ageing but this effect can be significant for class 2 capacitors.

## 4.6 Mica capacitors

Several different types of dielectric materials are used in mica capacitors, the most common being muscovite mica, the dielectric constant of which varies between 6 and 9. Mica is available naturally as silicates of aluminium with other chemical additives, and this can readily be laminated into thin sheets of thickness less than 0.025 mm. The material required for high stability mica capacitors must be very pure, and this is available in limited amounts as a natural resource. Relatively pure mica paper can also be produced artificially using commonly available forms of mica, mixing it with various organic materials and then putting it through a process similar to that used to make paper. This produces mica paper, called reconstituted mica, which has a laminated and rigid structure with a dielectric strength slightly less than pure mica. The dielectric strength of reconstituted mica can be increased by using impregnants but this limits the operating temperature range to between 100 °C and 400 °C depending on the type of impregnant used.

Mica capacitors can be made in a foil and film or in a metallised construction. In the metallised film capacitor, silver paste is screened onto the mica dielectric and this is then fired to form a bond between the silver electrodes and the mica dielectric. The individual mica capacitors can be stacked in series and parallel to obtain the required capacitance value. Metallised mica capacitors are also called silver mica capacitors, and they have a higher stability and tolerance than those of foil and film construction since there is closer control over the electrode thickness, and air spaces between the mica and the electrodes are minimised.

For high power radio frequency (RF), and high voltage applications a foil and film construction is used in which the mica is interleaved with tin–lead foil, and stacked to give the desired capacitance value. The mica stack is impregnated in vacuum with a wax having low electrical losses to remove air, and then clamped for added strength, and packaged.

Several packaging techniques are used such as moulded plastic, ceramic case and dipping in epoxy resin. The case protects the mica from moisture and gives it mechanical strength. For RF applications the inductance of the connections need to be minimised and a button case construction is then used. The mica stack is enclosed in a metal case in the shape of a button. One terminal of the capacitor is in its centre and the other terminal is the periphery of the button which is the metal case. This allows current to flow in a 360° arc from the centre terminal, which gives the shortest RF path and low self-inductance. The short thick terminal connections used also result in low external inductance. The mica capacitor is sealed in the metal case by welding or resin sealing.

Reconstituted mica paper can be wound with pure aluminium foils to form a different type of foil and film construction. The assembly is impregnated after winding and then compressed to give a solid capacitance block which can be packaged.

(Mica capacitor characteristics are summarised in fig. 4.9. They have very low RF loss and high stability so that they find use in RF circuits, and in applications such as oscillators and filters where stability is important.)

### 4.7 Capacitors for special applications

The capacitors to be described in this section are those used for power applications, and variable capacitors (also known as trimmer capacitors).

#### 4.7.1 *Power capacitors*

Capacitors for use in power applications need to withstand high voltage spikes and complex waveforms, and provide pulses of heavy current. They are sometimes used in transportable equipment, so that size and weight are important considerations, and they often operate in confined spaces so that heat dissipation can be a difficulty. The rating that a particular circuit demands of the capacitor can be determined approximately, but generally voltages and currents can only be found accurately by actually measuring their value with the capacitor in place. Capacitor case temperature rise is also measured as an indication of whether ratings are being exceeded. For power applications, the mounting for the capacitor must be carefully selected so as to minimise shock and vibration while maximising the heat dissipation from the capacitor case.

Power capacitors are generally made from paper or plastic film using the foil and film or metallised electrode construction. The heat generated in the capacitors is primarily due to current flow in the connecting leads and electrodes, and due to the dielectric loss. For many applications forced cooling of the capacitor is required. Film and foil capacitors can carry larger currents than metallised film versions, but they are also physically larger. If the peak circuit current is many times bigger than the rms current, the capacitor current rating is primarily determined by the bonds between the metal electrodes and the end terminals.

Power capacitors are often protected against overloads, one method being shown in fig. 4.12. The crimps in the case enable it to be forced

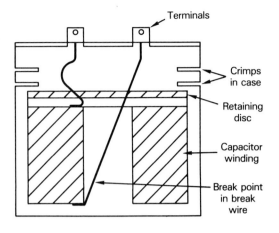

Fig. 4.12. Power capacitor with in-built overload protection.

out by about 10 mm under overload conditions and this causes a break in the wire connecting the terminal to the capacitor winding, isolating the winding.

#### 4.7.2 *Variable capacitors*

Referring to (4.1), it is seen that the capacitance value can in theory be altered by changing the dielectric constant, the spacing between the capacitor electrodes or the area of overlap between the electrodes. Changing dielectric constant is not feasible so all commercially available continuously variable capacitors alter the plate spacing or plate area.

Several electrical and mechanical parameters need to be considered in the choice of variable capacitors. The main electrical requirements are good temperature stability and high voltage operation, and a high $Q$ factor at high frequencies. Typical electrical characteristics for some of the more commonly available types of capacitors are shown in fig. 4.13. Mechanical parameters which need to be considered are ease and stability of adjustment, smooth torque, low contact noise, the fineness of adjustment measured in the number of turns to cover the full range, and the operating life measured in number of cycles. Generally trimmer capacitors are adjusted less than one hundred times during their life but where several hundred adjustments are required the shaft which moves the capacitor plates operates in ball bearings.

## 4.7 Capacitors for special applications

| Type | Capacitor range (pF) | Typical Q value at 20 MHz | Voltage (V) | Temperature coefficient (ppm/°C) |
|---|---|---|---|---|
| Air : open plate | 1 to 200 | 1500 | 800 | ±50 |
| Air : cylinder | 1 to 10 | 3000 | 300 | ±50 |
| Ceramic | 1 to 50 | 1000 | 100 | +1000 |
| Glass | 0.5 to 150 | 2000 | 1500 | ±50 |
| Plastic | 1 to 200 | 2000 | 1000 | ±50 |
| Mica | 1 to 5000 | 150 | 500 | ±200 |

Fig. 4.13. Typical characteristics of variable capacitors.

The air type variable capacitor is available in two forms, open plate and cylinder. In the open plate type the movable part (rotor) consists of several semi-circular discs on a shaft which can rotate. The static part (stator) has similar discs but is fixed such that the rotor and stator plates intermesh when the rotor is turned, and this effectively varies the overlap path area of the capacitor. Contact is made to the rotor via brushes or bearings. The other type of air capacitor is made from concentric cylinders and is a multiturn version. The stator cylinders are fixed at one end of an insulated tube and the rotor cylinders move into the tube at the other end via a screw drive. Adjusting the screw varies the overlap area between the two cylinders which operate as the capacitor electrodes.

The air capacitor has excellent stability and a low temperature coefficient of capacitance since the dielectric constant of air is relatively stable. The capacitor has good resistance against shock and vibration, high $Q$ and voltage rating and low inductance. It has the disadvantage of a maximum adjustment of 180° travel for the open plate type, and large size since the dielectric constant of air is 1.0.

In the ceramic variable capacitor the rotor consists of a semi-circular metallisation pattern which is formed onto an insulator by deposition, screening or painting. The stator consists of a second metallisation on a low loss ceramic base and the dielectric between the rotor and stator is ceramic. Movement of the rotor changes the plate overlap area and hence the capacitance value. The parts are held together by spring pressure to provide good contact between all areas, and so minimise the air spaces and give good electrical stability.

The ceramic capacitor is low cost, has a high capacitance–voltage ratio and a high Q value with low inductance. It is brittle and is not able to withstand mechanical shock and vibration. The maximum adjustment range is 180° so that fine tuning is difficult, it can be damaged by over-voltage and it has a high temperature coefficient of capacitance. It is not suitable for use in precision applications.

In the glass variable capacitor the stator consists of metallisation which is deposited, screened or painted onto a glass tube. The rotor is a piston which acts as the second metal electrode of the capacitor and is moved in and out of the stator by a threaded screw rod, so varying the plate overlap area. The glass capacitor can operate at a high voltage and has a smooth multiturn adjustment capability and a fairly linear capacitor–rotation curve. It is brittle and physically large and has a low capacitance value and range so that it is primarily used in applications where accurate trimming is needed.

The plastic dielectric variable capacitor uses semi-circular plate electrodes, similar to those used in air capacitors, but these are separated by a dielectric made from a plastic material. These capacitors have a higher capacitance value and voltage rating than air capacitors. They are also available in miniature form for use in hybrid circuits and chips of 5 mm by 2 mm are capable of having a value of about 10 pF, 300 V rating and a $Q$ value of about 1000.

Mica dielectric variable capacitors work on the compression principle to vary the capaci-

| Application | Capacitor type | Comments |
|---|---|---|
| High capacitance value. High *C/V* ratio. Charge storage, low frequency filtering | Electrolytic | Long life types for low ESR and high reliability |
| High stability. Counting, timing, filtering at high frequencies | Mica | Low *C/V* ratio |
| Moderate stability. Moderate *C/V* ratio. High voltage general purpose. Bypass, coupling, filtering at medium frequencies | Ceramic | Wide choice of parameters. Multilayer for high *C/V* ratio |
| Power applications at medium frequency. Alternating applications | Plastic film | Replacing paper |
| Power applications at low frequency. Power factor correction, motor start | Paper | Being replaced by plastic |

Fig. 4.14. Fixed capacitor selection guide.

tance value. They consist of a stack of mica plates which are inter-leaved between conducting metal plates, the alternate plates being connected at the two ends to form a stack of parallel capacitors. By this method large capacitance values can be obtained. The unit is sealed in a ceramic container and the capacitance value is changed by an adjusting screw which compresses the metal plates together and so varies their separation. The mica capacitor is low cost, has good stability and temperature coefficient, gives a smooth adjustment and multi-turns, has low inductance, a high capacitance–voltage ratio, and has moderate resistance to shock and vibration.

### 4.8 Capacitor selection

It is often difficult to select the right capacitor for any application primarily because of the large variety of capacitors available and the many trade-offs which are possible between them. Fig. 4.14 provides a very broad indication of the capacitor types which are used for the various applications and this should be used in conjunction with fig. 4.9 to ensure that the parameters meet the application requirements.

# 5. Magnetic components

## 5.1 Introduction

Components which are based on a magnetic effect cover a large number of different device types and technologies. Permanent magnets (section 5.5) are the simplest in construction. Transformers (section 5.3) and inductors (section 5.4) have an electrical coil in addition to a magnetic circuit and operate due to the electromagnetic effect. Relays (section 5.5) are mechanical components, whose output consists of the operation of moving contacts, but whose drive mechanism is a magnetic field created by an electrical coil. Hall effect devices (section 5.6) and *magneto resistors* (section 5.7) are semiconductors which are operated by a magnetic field.

## 5.2 Permanent magnets

### 5.2.1 *The demagnetisation curve*

The second quadrant of the magnetisation curve (see fig. 5.1) is the most important when considering permanent magnets. This part of the characteristic is called the demagnetisation curve. The area enclosed within the total magnetisation curve is a measure of the energy

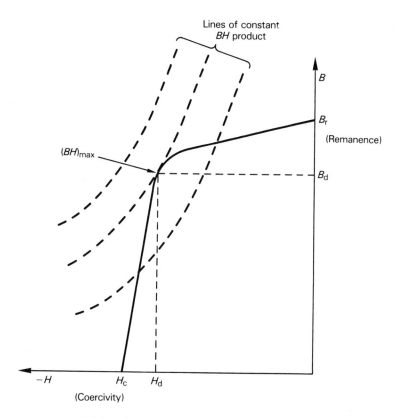

Fig. 5.1. The demagnetisation curve of a permanent magnet.

stored in the material and for permanent magnets this area should be as large as possible.

For a given value of flux density, the *BH* product gives the energy available in the material. Fig. 5.1 shows lines of constant *BH* product plotted on the demagnetisation characteristic and the line which just touches this curve is the maximum value of *BH*. The flux density $B_d$ and the field strength $H_d$ corresponding to $(BH)_{max}$ is the point at which a permanent magnet should be operated for most applications in order to get maximum efficiency.

The main parameters considered in the design of a magnetic circuit are the field strength required in the air gap (e.g. as in fig. 5.6), the dimensions of the air gap, the leakage factor, which is the ratio of the total flux to the useful flux, and the *reluctance* of the magnetic assembly. If $B_g$ is the flux density in the air gap in tesla, $l_g$ the length of the air gap in metres and $H_d$ the demagnetising field strength of the magnet in amperes per metre, then the effective magnetic length $l_m$ in metres required for the magnet is given by

$$l_m = 10^7 \, qB_g l_g / 4\pi H_d \qquad (5.1)$$

In (5.1), $q$ is the loss factor due to unwanted reluctances in series with the useful air gap. These reluctances are compensated for by taking $q$ to be about 1.1 so that the effective magnetic length is increased by about 10%.

The cross sectional area of the magnet, $A_m$, in square metres is given by

$$A_m = pB_g A_g / B_d \qquad (5.2)$$

where $B_d$ is the flux density in the magnet in teslas and $A_g$ is the area of the air gap in square metres.

Factor $p$ varies from one application to another between the values of 1.0 and 5.0.

A permanent magnet will lose some of its initial magnetisation due to changes in temperature, external magnetic fields, mechanical shock, vibration and time. The change with temperature may be reversible or irreversible. An irreversible loss of magnetisation occurs when the temperature is reduced, and this loss can only be restored by remagnetisation. No further loss of magnetisation occurs on subsequent exposure to this temperature.

At temperatures above the Curie point for the material, demagnetisation will occur although the material can be remagnetised after it cools. At very high temperatures, permanent metallurgical changes may occur in the material which degrade its permanent magnet performance. This temperature may be greater than or less than the Curie temperature depending on the type of magnetic material.

External applied fields can cause reversible or irreversible changes in magnetization depending on the magnitude of the external field. In order to prevent an irreversible change the magnet should be subjected to a stabilising field which has a strength greater than that expected from the external field.

Mechanical shock and vibration set up waves in the material which causes a loss of magnetism. This loss is small in modern magnetic materials, especially if they have been stabilised previously, either in an alternating magnetic field or by temperature cycling.

### 5.2.2 *Magnetic materials*

The ideal properties of a permanent magnet material are high *remanence,* and a large *coercivity* and energy product. No single material has all these properties. Higher coercivity materials allow the construction of short magnets with large cross sectional areas

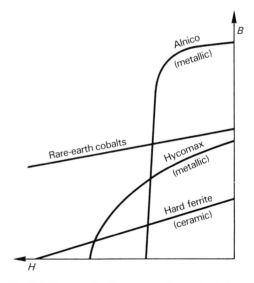

Fig. 5.2. Demagnetisation curves of some typical magnetic materials.

| Material | $B_r$ (T) | $H_c$ (kA/m) | $(BH)_{max}$ (kJ/m³) | Curie temperature (°C) | Temperature coefficient %/°C |
|---|---|---|---|---|---|
| Ferrite (ceramic) | 0.4 | 220 | 30 | 450 | 0.2 |
| Alnico (metallic) | 1.2 | 60 | 40 | 850 | 0.01 |
| Rare-earth cobalt | 0.8 | 1000 | 200 | 700 | 0.05 |
| Hycomax (metallic) | 0.8 | 100 | 45 | 850 | 0.01 |

Fig. 5.3. Typical properties of some magnetic materials.

that can be used without fear of substantial self-demagnetisation.

Fig. 5.2 shows the demagnetisation curves for some popular magnetic materials. Alnico has the highest remanence, which is two to three times that of a ferrite but its *coercive force* is less than half. Rare-earth cobalts have a high coercive force. They are expensive and are used when low weight and volume are required from a high performance magnet, such as in aerospace applications. Fig. 5.3 summarises typical parameters of a few magnetic materials. The rare-earth cobalts change in magnetisation very slowly at a temperature of about 100 °C, but at 200 °C the magnetism can change by 20% in a year, therefore these materials are not recommended for use in high temperature long life applications. For this type of application ceramic and metal magnets with low ageing rates are preferred.

## 5.3 Transformers

### 5.3.1 *Transformer parameters*

The peak secondary voltage of a transformer will change between the no-load and full-load currents due to voltage drops in the secondary and primary windings even if the primary voltage is kept constant. The *regulation* of the transformer defines this change and is usually stated as the change between the voltage ($V_N$) obtained on no-load current and the voltage ($V_F$) obtained on full-load current, measured as a percentage of $V_F$. The efficiency of the transformer is a measure of how well it converts the input power into the output. The excess of input power over output power is dissipated in the transformer as losses and generates heat. The power factor of the transformer is the ratio of the power input to the volt–ampere (VA) input.

Transformers often need to withstand high voltages between the secondary and primary windings. This is especially true in transformers where the primary is connected to the mains and the secondary voltage is low and accessible to a human operator. High voltages can result in *corona,* dielectric failure, surface creepage and flashover between points.

Corona is partial discharge within the transformer. It results in a.c. stress within the insulation system and can destroy it. Corona also generates radio frequency interference which can affect adjacent equipment and circuits. Corona increases with the magnitude of the applied voltage or frequency.

Flashover is arcing between parts of the transformer and creepage is flashover across the surface of the insulation. Both these result in high voltages across secondary circuits.

The dielectric strength of the insulation between the primary and secondary windings is usually measured as a maximum withstanding voltage per unit thickness of insulation. Solids have a higher dielectric strength than liquids and gases. In a transformer, the presence of gas adjacent to a solid insulator presents a weakness in which corona can be generated, and this limits the maximum voltage which can be applied across the insulation system. It is therefore important to avoid air gaps in series with the insulation. When several insulating materials are used in series the stress in each is inversely proportional to its dielectric constant. Therefore the insulation with the lowest dielectric constant has the highest stress, and this is usually air or gas. When a direct voltage is applied to the material the voltage drop is mainly due to its resistivity so the material with the highest resistivity has the highest stress.

Transformers can be *shielded* electrostatically

and electromagnetically. Electrostatic shielding reduces the voltage transfer through the interwinding capacitances. It is needed to prevent the transfer of transient voltages or high frequency noise from the power input circuit to the secondary circuit. The shield is usually a grounded metallic plate between the primary and secondary windings. Electromagnetic shielding is used to attenuate the magnetic field which leaks from the magnetic circuit of the transformer. These fields may induce voltages in adjoining circuits. Placing a magnetic shield around the transformer is not usually very effective since most of the stray lines of flux from the transformer would be perpendicular to the shield and would pass through it. It is much better to separate the transformer and adjoining sensitive circuits and to orient them to minimise pick up. The adjoining circuits can also be shielded by layers of thin high permeability material which are usually interleaved with layers of a non-magnetic material such as copper.

### 5.3.2 Types of transformers

Power transformer cores are generally made from thin sheets of iron with nickel or cobalt additives. These are cut into laminations which are insulated from each other to reduce eddy currents. These cores are generally used for frequencies up to about 50 kHz. For higher frequencies, up to 300 MHz, ferrite cores are used. These can have a toroid or pot structure similiar to that shown in fig. 5.6. All magnetic cores generate acoustic noise due to the magnetostriction effects under cyclic flux changes. This core noise is most noticeable at high frequency.

Temperature rises resulting from the VA requirements of large power transformers determine their physical size. In a small transformer the physical size is determined by the voltage drop requirements. Therefore, a large transformer is thermally limited whereas a small transformer is impedance limited. The overall dimensions of a power transformer are approximately proportional to $(VA)^{1/4}$ and the weight and losses to $(VA)^{3/4}$.

Transformers used in switching converters operate at frequencies up to several hundred kilohertz. Most feedback type converters need a sharp saturation knee, i.e. a square $BH$ loop, with low residual inductance in the saturated state, so that the switching losses are minimised and the slope of the waveform and voltage overshoots are reduced. Materials used for the cores are toroids with 50% nickel and 50% iron for frequencies up to about 50 kHz, and ferrites for higher frequencies.

Pulse transformers must be capable of passing a square wave or a pulse having a short rise and fall time without appreciable distortion of the waveform. Fig. 5.4 shows a typical output voltage waveform from a pulse transformer and indicates the terms used to describe it.

A pulse transformer should have small size, a short rise time, a good pulse width to pulse rise time ratio (called the span ratio) and be capable of resolving adjacent pulses in a high pulse repetitive frequency application. Transformers can be built having a small size, and several devices can be mounted inside a dual-in-line package. Usually these small transformers have a primary inductance less than about 2 mH and a maximum voltage time product of about 10 V μs. To get a good span ratio requires a large core size, and to resolve adjacent pulses the positive and negative halves of the waveform must have equal voltage-time products. This usually places a requirement for a high backswing although it is usual to limit this in circuit applications as it adds to the rating of semiconductors which may be driving the pulse transformer. The primary inductance will be high if a large step-up ratio is used in the transformer, and this should be avoided if a high bandwidth and span ratio are required.

A current transformer is designed to give a low power output signal which is proportional to one or more alternating current inputs. It is used to measure the power and current in a circuit, to detect threshold currents and to sample current waveforms for instruments. The voltage-ampere product of the load at the rated secondary current is referred to as the burden of the current transformer, and the transfer ratio is the ratio of maximum input current to burden current.

For current transformers designed for low input currents the primary and secondary windings are wound as coils on the same magnetic core. For large primary currents the centre of the core is open and the cable carrying the current is looped once or twice around

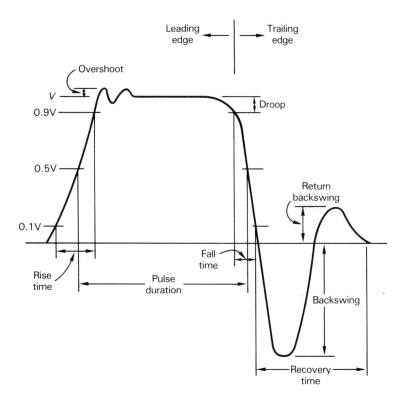

Fig. 5.4. Output from a pulse transformer.

the core. Generally a toroidal core is used which may be made from grain oriented silicon steel, nickel-iron alloy, molybdenum-permalloy or Mumetal. The highest accuracy is obtained from core materials having the lowest magnetic strength, $H$. The burden and accuracy determine the physical size of the transformer and the core must be large enough to accommodate the primary and secondary turns and the insulation.

If the transformer is used to measure currents only, it is important that the secondary current is a known fixed ratio of the primary current. In practice the current ratio differs from the turns ratio between primary and secondary windings by a value which depends on the magnitude of the transformer excitation current and the secondary circuit current and power factor. Therefore, the current ratio is not constant but varies with frequency and load. The amount by which the current ratio differs from a constant value is known as the magnitude error of the current transformer.

The secondary current is displaced by almost 180° from the primary current. If this angle is exactly 180° then no error would be introduced when the current transformer is used with a wattmeter to measure power. The phase error of the current transformer is a measure of the amount by which the angle between the primary and secondary currents differs from 180°.

An *autotransformer* consists of a single coil which is tapped to provide the primary and secondary windings. The tap ratio of the transformer ($r$) is the ratio of the secondary to primary voltage. The advantage of an autotransformer compared to a two winding transformer is smaller physical size, although there is very little size advantage for small values of primary to secondary turns ratio below about 0.5. An autotransformer also has less regulation and a lower leakage inductance. The disadvantage of an autotransformer is that it does not provide electrical isolation between primary and secondary.

## 5.4 Inductors

In its simplest form an inductor consists of a coil of wire into which a magnetic core of permeability $\mu_c$ is introduced. If $l$ is the mean magnet length, $a$ the cross sectional area of the core and $n$ the number of turns of the coil, the inductance $L_c$ is given by

$$L_c = \mu_c a n^2 / l \quad (5.3)$$

$L_c$ will have units of henries if $\mu_c$ is in henries per metre, $a$ in square metres, and $l$ in metres.

Due to the shape of the $BH$ curve the permeability is not constant, but varies with the flux density in the core. Therefore, magnetic cored inductors have a linear inductance over a limited current range only. Better linearity can be obtained by introducing an air gap into the core as shown in fig. 5.6. This will give an effective permeability of $\mu_e$ where the ratio $\mu_e/\mu_c$ is typically 0.1.

The air gap in the inductor core also reduces the effect of the spread in value of $\mu_c$. A ±20% tolerance on material is reduced to about ±2% with an air gap. The temperature coefficient of inductance is also improved by introducing the air gap in the core. If $T_c$ is the coefficient of the core material and $T_e$ is the effective coefficient with an air gap, then over a temperature range $\Delta t$

$$T_c = \Delta \mu_c / \mu_c \, \Delta t \quad (5.4)$$

$$T_e = \Delta \mu_e / \mu_e \, \Delta t \quad (5.5)$$

$$T_e = T_c \, \mu_e / \mu_c \quad (5.6)$$

$$= (\Delta \mu_c / \mu_c^2 \, \Delta t) \mu_e \quad (5.7)$$

$$= F \mu_e \quad (5.8)$$

$F$ is called the temperature factor of the material.

Although the air gap introduces greater stability into the operation of the inductor it also reduces the inductance value of $L_c$ to $L_e$ where

$$L_e = L_c \, \mu_e / \mu_c \quad (5.9)$$

An important consideration in the design of an inductor is its $Q$ factor. This is given by the ratio of the reactance of the coil to the effective resistance of the inductor

$$Q = X_L / R_e \quad (5.10)$$

$R_e$ includes the winding resistance and the equivalent core loss resistance. In an ideal inductor $R_e$ is zero and $Q$ is infinite. The smaller the value of $Q$ the higher the losses in the inductor.

The $Q$ factor can also be stated as the reciprocal of the loss tangents in an inductor:

$$Q = 1/\tan \delta \quad (5.11)$$

The losses involved can be itemised as:
(i) Loss due to the d.c. resistance of the winding. This is the major loss at low frequencies.
(ii) Loss in the winding due to *skin effect* at high frequencies, which causes the current to flow near the surface of the conductor so increasing its d.c. resistance.
(iii) Loss in the winding due to the proximity effect. The magnetic field is perpendicular to each winding and produces eddy currents in them which oppose the field. This causes major losses at high frequencies and Litz wire should be used to reduce it.
(iv) Loss due to the stray capacitance of the winding. This loss increases rapidly with frequency.
(v) Residual, eddy current and hysteresis losses in the core.

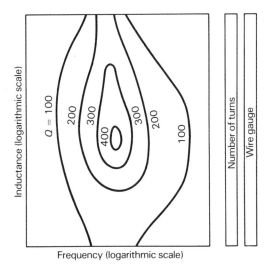

Fig. 5.5. Iso-$Q$ curves for an inductor.

Because of the importance of the $Q$ factor and its variation with different operating conditions inductor data sheets usually provide iso-$Q$ curves, as shown in fig. 5.5, so that for any combination of the inductance, number of turns and wire gauge the designer can read off the $Q$ factor of the coil.

Another factor which is frequently used in design calculations is the $A_L$ of the coil where

$$A_L = L/n^2 \qquad (5 12)$$

From (5.3) it is seen that the $A_L$ value should remain constant for a given core, so that if this value is given in the data sheets the number of turns for any value of inductance can be readily obtained. However, the $A_L$ figure should be used with caution. It is usually defined at low flux density and inductances increase with flux density. Also, if an air gap is used flux fringing will occur near this gap so that not all the turns in the coil will contribute the same inductance. The inductance will therefore depend on how a partly wound bobbin is filled, and the $A_L$ values given in the data sheets assume a fully wound bobbin.

Inductors are used in a wide range of applications from current limiting and tuning to energy storage and filtering. Alternating current reactors are used for current limiting, for simulating specific power factor loads or for power factor correction of capacitive loads. The inductors must have stability of inductance with changes in voltage and current and must be able to handle large voltages without saturation. Usually, silicon–iron laminations or strip wound cores are used, with an air gap in the magnetic circuit. At higher frequencies, above about 10 kHz, ferrite cores are preferred.

Smoothing chokes are used in d.c. power supply applications to provide a high impedance to the a.c. ripple, and low impedance to d.c. power. Close tolerances are generally not needed and an air gap is essential to prevent saturation of the core by the direct current.

Inductors used for filtering applications must have a close tolerance on the inductance value, a high stability to temperature and a.c. level changes, and a high $Q$ value. For low frequency use the core may be made from low loss nickel–iron laminations, and at higher frequencies ferrite or powdered nickel–iron cores are more common. The powdered cores

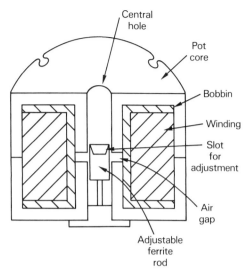

Fig. 5.6. Ferrite pot core with adjustable inductance (half section).

are usually toroidal in shape and the inductance value is adjusted by altering the number of turns. Ferrite materials are usually made into pot cores and the inductance value can be finely adjusted after assembly by moving a small ferrite rod in the region of the air gap to vary the effective gap. This arrangement is shown in fig. 5.6.

## 5.5 Relays

### 5.5.1 *Relay parameters*

A relay is a component which uses an electromagnetic effect to mechanically open and close sets of *contacts* which carry current. Fig. 5.7 shows a simplified diagram of one form of

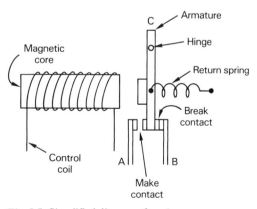

Fig. 5.7. Simplified diagram of a relay.

relay. When there is no coil current the return spring pulls the armature back against the break contacts so that an electrical circuit exists from C to B. When sufficient coil current is passed it energises the magnetic core and the armature is pulled towards it, against the force of the return spring. This causes the break contact to open and the make contact to close so that an electrical current can pass between terminals C and A.

The power required to energise the control coil is usually very small but the contacts can make and break a greater amount of power so that there is gain in the relay between the input and output circuits. The relay also provides electrical isolation between the input control coil and the output contacts.

Fig. 5.8 shows the symbol for a relay. Only one contact is shown here although practical relays can have tens of contacts all controlled by a single coil. These contacts can be arranged to operate in various sequences.

Fig. 5.9 shows the symbols for a few contact arrangements. Note the alternative terms which may be used to describe them. Generally the form letters are used so that relays can be referred to as 1A, 2B and so on.

The main parameters to be considered during the circuit design stage are:

(i) the coil voltage and current required to operate the relay;
(ii) the rating of the relay contacts;
(iii) the time required for the relay to switch on and off;

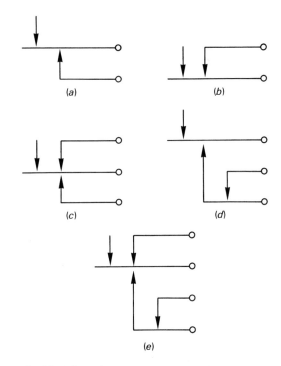

Fig. 5.9. A few relay contact arrangements; (a) form A, single pole, single throw, make, (b) form B, single pole, single throw, break, (c) form C, single pole, double throw or changeover, break before make, (d) form D, single pole, double throw, make before break, (e) form E, single pole, double throw, break, and make before break.

(iv) the dielectric strength of the insulation between the input control coil and the output load contacts.

Fig. 5.10 shows simplified waveforms for a relay with a single normally open contact. The voltage is applied across the relay coil at time $A$ and the coil current builds up exponentially towards the d.c. current. At point $B$ the relay is energised and its contacts close. Time $AB$ is referred to as the *operate time* of the relay. When the relay contacts close they *bounce* open again and after a period known as the *bounce time* the contacts finally remain closed. Thereafter there is still some variation in *contact resistance* for a period although the contacts do not open, and this is known as relay *chatter*.

Relays are available for operation with a.c. or d.c. coil currents. Direct current relays are more common since they can run from control

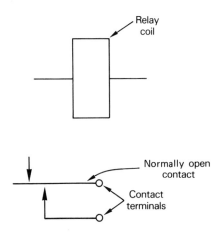

Fig. 5.8. Relay symbol.

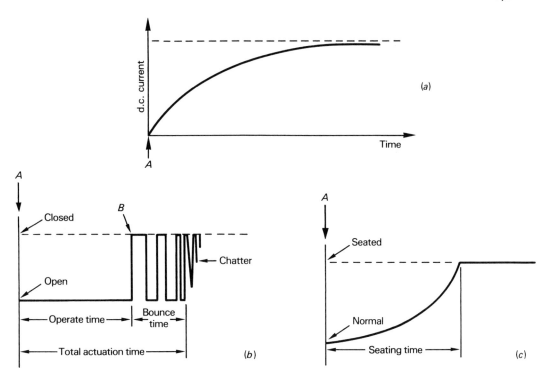

Fig. 5.10. Relay operation sequences. (a) Coil current. (b) Contact position. (c) Armature position.

circuits and interface more readily to other electronic circuitry. Direct current relays also have longer life since there is absence of a.c. vibrations during the closing periods. Because the tendency to chatter is less on d.c. relays they can also be designed to operate with lower coil currents and lighter pull-in forces, so that a d.c. relay usually has higher sensitivity than an a.c. relay. The power loss in the magnetic circuit is lower in a d.c. relay, and since a lower holding force is needed the copper loss is also less, so that a d.c. relay can be made more efficient than an a.c. relay. Direct current relays are generally cheaper and can work over a wider voltage range than a.c. relays.

There is a delay in the operate and release times of a relay due to the time required for the coil current to change. Generally these times can be made very short by careful relay design. For long time delays, the relay can be designed with a copper slug as shown in fig. 5.11. For a delay in the operating time the copper slug is placed at the armature end. This acts as a short-circuited coil and retards the build-up of the magnetic field when the

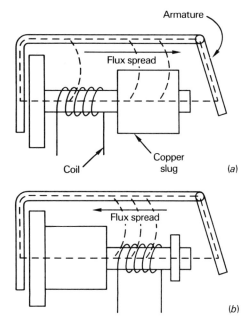

Fig. 5.11. Increasing relay switching time with a copper slug; (a) delayed operate time, (b) delayed release time.

coil is energised. For a delay in the release time the copper slug is placed at the heel end of the relay and this holds the armature in until the flux has decayed to a very low value. If delays are needed in both the operate and release times then a full length copper sleeve can be used along the core of the coil.

Design trade-offs are required when short operate and release times are needed in a relay. The operate time can be reduced by overdriving the coil, reducing the coil time constant, reducing the force on the contacts and the armature return spring, and reducing the armature travel. The release time is reduced by increasing the force on the armature return spring, by reducing armature travel and by ensuring a rapid decay of current in the coil i.e. adopting no protective circuits across the coil to prevent voltage overshoots which can damage other circuit components.

The sensitivity of the coil should be as high as possible so that coil drive requirements are reduced, but here again compromises are needed. The sensitivity can be increased by decreasing the armature travel and reducing the armature spring return force. This, however, makes the relay less resistant to shock and vibration. Sensitivity can also be increased by reducing the contact force but now the relay cannot switch large currents, and the reliability of the relay when carrying low voltages and currents is also reduced.

### 5.5.2 Relay contacts

Relay contacts fail due to several reasons and these include:

(i) Corrosion of the contacts due to contaminants in the atmosphere, such as sulphur, chlorine and moisture. This can be protected against by enclosing the relay in a sealed case. However contamination can also occur due to gases released from the insulation inside the case during relay operation.

(ii) Erosion of the contacts caused by arcing, welding and glow discharge. In the case of a d.c. load this would result in metal transfer from one contact to another. For a.c. loads, metal transfer does not occur but there is net loss of metal due to evaporation.

(iii) Welding of contacts. Spot welding can occur due to the heating effect of the current at the point of contact, and cold welding results when the surfaces are cold but under pressure. Tearing of the weld causes loss of metal from the contacts.

On inductive loads relay contacts need to be derated to about half the value used for resistive loads because of contact arcing. When the load is inductive the energy stored in the load is dissipated as an arc across the contacts when they open unless external protection circuits are used.

Current surges during contact closing can cause damage since the contact pressure is light and the contacts slide and bounce. A small weld occurs at the point of contact closure and in d.c. circuits this breaks and gives metal transfer. To prevent this heavy duty contacts and high contact force are needed to reduce bounce. The contact material must also have high electrical and thermal conduction.

When contacts operate at very low currents and voltages they are said to be running 'dry' circuits. The contact resistance in these conditions can be very high, up to many ohms, and contact failure occurs more often from over rating contacts than from under rating them. Each contact has a minimum current and voltage for its material, shape and size. Mechanical pressure and the current and voltage must be sufficient to cause slight melting and fusing of the contact surfaces at each closure. Pressure is also needed to remove surface films, and the contact current density is determined by the contact current and the shape of the contact mating surfaces.

Fig. 5.12 gives the minimum voltage required across various contact materials in order to cause them to soften and give a reliable join. The contacts will initially meet on any irregularities on their surface and this will result in high contact resistance. If the contact voltage is above the softening value then the bumps will soften, since they carry a high current density and generate heat faster than it can be carried away. This softens the material at the contact points and increases the contact surface area so that the contact resistance is decreased. At larger contact voltages melting and vaporisation can occur at the contact areas after closure.

Gold is used as a contact material for low currents, below about 100 mA and voltages below 30 V. Pure gold is not suitable as it welds

| Contact material | Softening | | Melting | |
| --- | --- | --- | --- | --- |
| | Voltage (V) | Temperature (°C) | Voltage (V) | Temperature (°C) |
| Silver | 0.09 | 180 | 0.37 | 960 |
| Gold | 0.08 | 100 | 0.43 | 1063 |
| Palladium | 0.20 | 450 | 0.57 | 1500 |
| Tungsten | 0.40 | 1000 | 1.00 | 3400 |

Fig. 5.12. Softening and melting characteristics of some common contact materials.

even in the absence of current, and although impurities are added for greater hardness the mechanical life is still short. Gold plating is, however, often used on relays required to operate in dry circuits.

Silver and silver alloys give good general purpose relay contact materials. Silver tarnishes easily but this film can be disrupted at relatively low voltages. For this reason silver contacts are not suitable for use below about 50 V or on light loads. Cadmium is often added to silver for heavy duty contacts since it is better able to resist arcing and metal transfer on d.c. loads. Palladium alloys and palladium are used for large inductive loads.

### 5.5.3 Moving armature relays

The types of relays covered under the definition of moving armature are diverse, and range from large power relays to small TO-5 relays intended for printed circuit board mounting. They all operate on a similar principle where the coil attracts the armature directly and the moving contacts of the relay are controlled by the armature.

Fig. 5.13 shows the construction of a general purpose relay which, because of its construction, is also referred to as a 'clapper' type relay. The armature is hinged on the U-shaped heelpiece although L-shaped heelpieces are also in common use. When the armature is released its return spring causes it to rest with its moving contact against one of the fixed contacts. When the coil is energised it attracts the armature, against the force of the spring, and causes it to move so as to make with the second contact. Although only one set of contacts is shown in fig. 5.13 a large number can be operated by the same armature.

General purpose and power relays using the construction shown in fig. 5.13 can operate at currents of 2 A to 25 A and more. They are relatively cheap and readily available, but they usually have a low life, are not resistant to shock and vibration and will not operate unless mounted in a fixed number of positions i.e. they are position sensitive. They are used in household appliances, coin operated machines, heating and lighting control, and in motor control.

Power relays are generally physically larger and more rugged than general purpose relays, and they also have thicker insulation and larger terminals. They are designed with safety in mind and usually have minimum creepage and clearance distances between the power and control circuits, carefully specified to meet the requirements of safety authorities in various

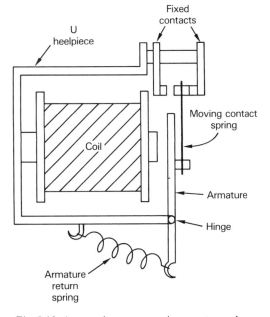

Fig. 5.13. A general purpose moving armature relay.

countries. The contacts are heavy duty to handle large inductive loads, and the contact gap and armature strokes are made wide, so that this type of relay has relatively low sensitivity.

Telephone type relays have been improved over many decades. They are usually designed to operate over a wide range of loads from low voltages carrying speech signals to the coil currents of other relays and electromagnets. This property also makes them useful for applications other than in telephone exchanges. The relays have a long slim design so that high mounting density is possible, and they have a large number, combination and variety of contacts all operated by the same armature. The relays have good density and contact reliability and allow good control over parameters such as timing. They are small and light, are largely insensitive to mounting position and will withstand moderate shock and vibration. Due to the close spacing between contacts they are difficult to service and are not suitable for power switching.

Crystal can and TO-5 relays are hermetically sealed and are therefore suitable for use in adverse environments such as demanded by military and aerospace application. They are small and light and are resistant to shock and vibration. The relays are designed with small contacts and low contact pressures so that they are suitable for switching low power loads only. They are also relatively expensive. The TO-5 relay is mounted on printed circuit boards along with other electronic circuitry. It is usual in this type of relay to have coil suppressor diodes or transistor coil drivers mounted inside the TO-5 can of the relay.

### 5.5.4 Dry reed relays

The reed relay is a telephone industry designed product which has subsequently found use in a wide variety of applications. Fig. 5.14 shows the basic construction of this relay and it is seen to consist of a central sealed reed capsule and an outer control coil. The coil flux acts directly on the reed contacts and no intermediate moving armature is involved.

The reeds are generally long and slender and are made from 'springy' material. When the coil is unenergised the reeds spring apart and the contacts are open. When the coil is excited the

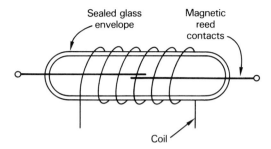

Fig. 5.14. Simplified diagram of a reed relay.

reeds form part of the flux path generated by the coil. This causes them to be magnetised and they are pulled into contact with each other so that an electrical circuit is made through them. It is possible to have normally closed reeds, biased by a permanent magnet into the closed position. The coil now generates a flux which neutralises the permanent magnet field in the reeds and opens the relay. Reed relays are also available in form C (see fig. 5.9) (change-over) and several reed envelopes can be operated in the same coil so that a variety of contact arrangements are possible.

Fig. 5.15 shows the equivalent circuit of a reed. $L_1$ is the inductance of the long thin

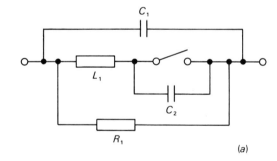

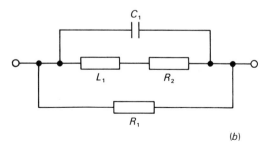

Fig. 5.15. Equivalent circuit of a reed; (a) reed switch open, (b) reed switch closed.

reeds, and $R_1$ is the insulation resistance across the glass envelope. The reeds also have a capacitance $C_2$ across the overlapping contacts although this is small. Capacitance $C_1$ between the long reed blades and the relay coil is usually much larger, of the order of 1 to 5 pF, and when several reeds are operated from the same coil their individual capacitances are connected in parallel. This capacitance effect can be reduced by placing an electrostatic shield between the reed switch and the coil. The reeds are also susceptible to interference from external magnetic fields so that magnetic shielding may be necessary.

Reed relays are manufactured in many sizes and assemblies. Special designs are required for high voltages (10 000 V) by wider terminal spacing, and for power handling (100-400 VA). High insulation resistances, greater than $10^{12}$ Ω are possible by careful control to reduce shunting paths.

Noise voltages are generated in reed relays in several ways. Because of the cantilever action of the long slender reeds they have a bounce time of about 0.5 ms and after this they continue to oscillate for a period of a few milliseconds during which time the contact resistance varies. Noise is also caused by the voltage induced in the reeds soon after closure due to the magnetostriction effect i.e. the force inducing a voltage in the material. The movement of the reeds in the magnetic field of the coil also induces in them a further noise voltage, and thermal voltages are generated in all relays due to the difference in temperature gradient across the metal contact junctions. Generally only thermal noise is important unless the reed relay is to be used to take readings within a few milliseconds of closing. Thermal noise can be minimised by special contact materials which conduct the heat and reduce temperature gradients, and by assembly techniques which reduce the coil power and therefore the generated heat. Thermal noise is typically of the order of 100 μV in reed relays.

The sensitivity of a reed relay is dependent on the sensitivity of the individual reed switches, on the number of reeds in the coil, on whether permanent magnet bias is used, and on the efficiency of the coil flux coupling to the reeds. The relay operating times depend on the coil time constant, the reed characteristic and the amount of coil overdrive. The release times given in data sheets are normally for unsuppressed relays and will increase if diode or RC suppressors are used across the coil. Reed relays can stand considerable vibration, of 20g or more, provided the frequency of vibration is different from the resonant frequency of the reeds. This resonant frequency varies from 1 kHz to 5 kHz depending on the reed size.

A number of different materials are used for the reed contacts, e.g. gold, silver, tungsten and rhodium. The reed material is usually nickel-iron which has good magnetic performance and resonant characteristics which minimise bounce. However this material is not an ideal electrical conductor so for some applications additives or other platings are used.

### 5.5.5 *Diaphragm relay*

In the diaphragm relay a thin metallic diaphragm is drawn by an electromagnet and acts as the moving contact. The relay consists of two parts, a sealed capsule containing the contacts and an external electromagnet. There is no moving armature and the relay is therefore similar to a reed relay.

Fig. 5.16 shows the cross section through a diaphragm relay. The central metallic rod has its ends electroplated to form a ring which acts as the fixed contact. The rod extends into the coil so that when it is energised it exerts a central pull on the diaphragm, but the contact between the diaphragm and fixed contact occurs around the central area so that the high spots are removed from the centre of contact force. The diaphragm can also tilt and adjust itself to fit snugly onto the fixed contact so that the presence of dirt on any spot is not serious.

The diaphragm is illustrated in fig. 5.17 and is usually made as a gold plated nickel-iron disc. It consists of a central contact area, a slotted area which forms three cantilever springs, and the outer mounting edge which is clamped in the relay assembly as shown in fig. 5.16. The diaphragm cover is welded in an inert atmosphere so that the diaphragm assembly is sealed, as in the reed relay. The magnetic circuit with the diaphragm cover is

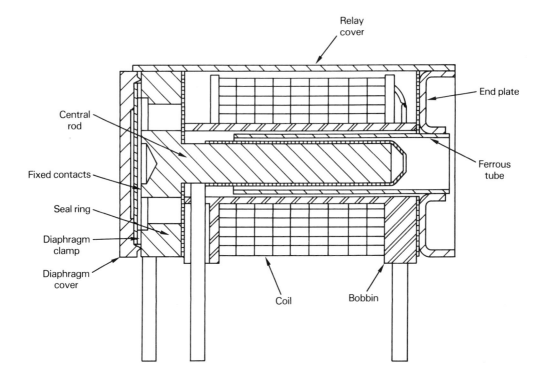

Fig. 5.16. The diaphragm relay.

completed through the seal ring, relay cover, end plate, and ferrous tube.

For its size the diaphragm relay has a very efficient magnetic circuit. A parallel flux path exists through the diaphragm and the ferrous cap giving high flux densities so the relay can operate with high contact forces. In a reed relay there is low magnetic efficiency since a compromise is required between the reeds having a large cross sectional area in order to carry high flux, and being slender and flexible to provide mechanical restraining forces. Reed relays therefore operate with small contact clearances and low contact forces.

The diaphragm relay has a high resistance to shock and vibration since the mass of the moving element is small, high contact forces are attainable, and the diaphragm cannot resonate like a reed. Generally the relay has negligible contact bounce but there is a short settling period when the contacts first close during which the contact resistance is higher than its final value.

### 5.5.6 *Mercury-wetted relay*

Fig. 5.18 shows the arrangement of a mercury-wetted relay. The armature is spring biased to one side and is deflected to the other contact (normally open) by a magnetic field. The mercury from the pool at the bottom is drawn up the armature and constantly wets the mating contacts. Electrical contact is therefore between mercury and mercury and the contact faces are constantly renewed. Because of this the mercury-wetted relay can switch a wide range of load currents without any detrimental

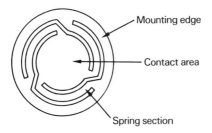

Fig. 5.17. Illustration of the diaphragm used in the diaphragm relay.

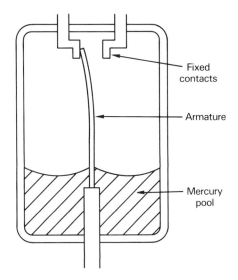

Fig. 5.18. Contact arrangement in a mercury-wetted relay (coil not shown).

effect on contact life. The mercury film also damps the armature movement so that the relay does not exhibit any significant contact bounce or chatter.

The contact resistance of a mercury-wetted relay is very low, in the region of 30 m$\Omega$, and this varies by less than $\pm 2$ m$\Omega$ during the life of the relay. This makes the relay very suitable for use in signal logging and routing applications. Contact noise is of the order of 30 $\mu$V to 75 $\mu$V. The disadvantages of the mercury-wetted relay are that it is relatively expensive, and that it cannot operate at low temperatures since mercury solidifies at $-30.8\,°$C. The relay is also position sensitive and must be mounted right side up and not more than about 25° away from the vertical.

There are many variations of the mercury-wetted relay principle. Reed relays have been built in which a film of mercury is used to stabilise contact resistance and increase life. Also in a form of power relay both contacts are fixed and conduction occurs by submerging them into a pool of mercury.

### 5.5.7 Thermal relays

Metals expand when heated and this property has been used in thermal relays, the expansion causing the movement of the contacts. The current flowing in the relay coil is used to heat the armature. These relays have long time delays which depend on the residual heat in the coil from a previous operation. It is also difficult to adjust the relays to operate accurately at a given temperature.

Recently a new type of thermal relay, called the thermomagnetic switch, has been developed which depends for its operation on the decrease in magnetic characteristics with temperature of some types of materials, notably alloys of ferrites. At the Curie temperature the permeability decreases rapidly and this occurs over a band of about three centigrade degrees. So over this temperature range the material switches from being magnetic to non-magnetic. These materials are called compensators.

Two arrangements for a thermal relay are shown in fig. 5.19. The relay coil (not shown) is wound around the compensator material and heats it.

The arrangement shown in fig. 5.19a and b is normally closed at low temperatures. At a temperature below the Curie point of the compensator material the flux path is through the material and the reed, so that the reed switch is closed. When the switching temperature of the compensator material is reached it becomes a paramagnet and inhibits flux flow from the magnet. This results in the flux through the reed being reduced and it opens.

The arrangement shown in fig. 5.19c and d is usually open at low temperatures, the flux from the magnet flowing through the compensator material so that the reed is open. When the temperature reaches the Curie point of the material it presents a high impedance to the magnetic flux so that this now flows through the reed closing its contacts.

In the thermal relays described so far the switching temperature is determined within close tolerances by the characteristics of the magnetic material, and this cannot be adjusted once the relay has been built. Many applications require that the switching temperature is adjustable although it does not need to be set accurately. These relays often use nickel-iron compensator alloys which have relatively linear characteristics. A relay arrangement using this is shown in fig. 5.20. At low temperatures flux $\phi_2$ is small, and at high temperatures flux $\phi_1$ is reduced. The change in flux through the com-

98  *Magnetic components*

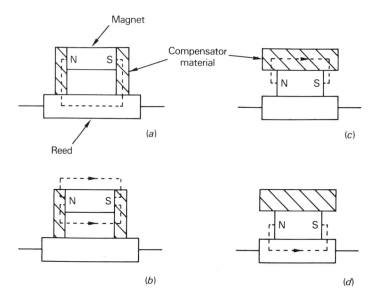

Fig. 5.19. Arrangements of a thermomagnetic relay; (*a*) break contact, low temperature, (*b*) break contact, high temperature, (*c*) make contact, low temperature, (*d*) make contact, high temperature.

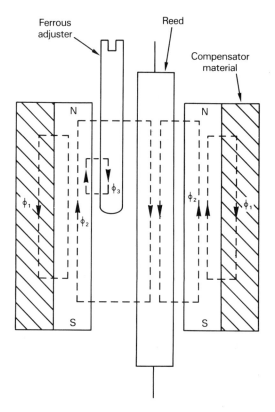

Fig. 5.20. Flux map of a thermomagnetic relay.

pensator is linear with temperature and the switching point of the reed relay can be adjusted over wide limits by screwing the ferrous adjuster in or out since this determines the fraction of the flux $\phi_3$ which is diverted from the relay.

The thermomagnetic switch can be used without a coil, as a temperature sensor, or it can be used with a heating coil as a thermal relay. When used with a coil it has a much longer operating time delay than a traditional electromagnetic relay, but it has the advantage that it is not position sensitive or affected by stray magnetic fields, and that it can operate with an alternating or direct current in the coil without any contact chatter.

### 5.5.8 *Hybrid and solid state relays*

A hybrid relay contains solid state and electromagnetic elements in the same package. The solid state circuitry may form the input or the output stage of the combination. Fig. 5.21*a* shows an arrangement where a relay is used as the input and the triac as the output stage. When the relay is energised it provides gate current to the triac so that it turns on. Solid state relays do not contain any electromechanical components.

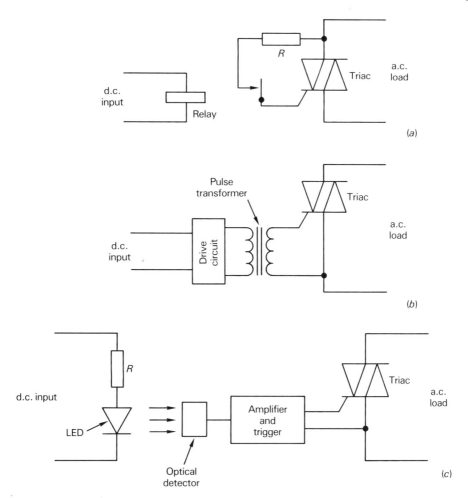

Fig. 5.21. Examples of hybrid and solid state relay arrangements; (a) hybrid with electromagnetic input, state with transformer coupling, (c) solid state with optical coupling.

The solid state components most commonly used as input devices are the LED and transistor. Output devices are usually power oriented devices such as transistors, thyristors or triacs. Isolation between input and output is by a relay in a hybrid circuit, or by a transformer or optical coupler in a solid state circuit. Fig. 5.21b shows transformer isolation and fig. 5.21c shows optical isolation.

A few of the parameters of solid state relays are summarised in fig. 5.22 and compared to those of other types of relays. The life of the solid state part of the relay is very long, and it also has a much higher reliability than the mechanical part. The electrical isolation between input and output of a solid state relay is as good as that obtainable from an electromagnetic relay. However when a solid state device is used as the power stage of a relay the leakage current through it is relatively high compared to that of an open mechanical contact. This leakage is also time and temperature dependent.

When a solid state device is conducting there is a voltage drop across it which varies only slightly with the magnitude of the load current. This drop is high compared to that across the contact resistance of an electromagnetic relay so that a solid state relay usually needs more cooling and is physically larger than its mechanical counterpart.

Electromagnetic relays can also have many

100  *Magnetic components*

| Parameter | Moving armature | Dry reed | Mercury-wetted | Solid state |
|---|---|---|---|---|
| Typical coil power (W) | 0.1 to 10 | 0.1 to 5 | 0.1 to 1 | 0.05 to 0.5 |
| Operating time (ms) | 2 to 15 | 0.2 to 1 | 1 to 4 | 0.1 to 2 |
| Release time (ms) | 1 to 15 | 0.01 to 0.5 | 1 to 3 | 1 to 10 |
| Typical load current (A) | 1 to 30 | 0.1 to 1 | 1 to 5 | 1 to 10 |
| Contact resistance (mΩ) | 50 to 100 | 40 to 250 | 10 to 50 | |
| Electrical life (operations) | $10^5$ | $10^8$ | $10^{10}$ | |

Fig. 5.22. Summary of relay parameters.

more contact arrangements on the same relay than a solid state output stage, and these are less susceptible to damage caused by overloads from current and voltage transients.

The advantage of solid state and hybrid relays is their flexibility. Solid state circuits can be incorporated to provide a variety of functions. For example, a class of relays, called time delay relays, use an electromagnetic output stage and internal solid state circuitry to give variable time delays between input and output. Solid state relays have no contact bounce, and their operating time can be made very short so that they can be switched at the zero voltage of an a.c. waveform. This reduces the amount of radio frequency interference which is generated during power switching. If the solid state device is a triac or thyristor operating from 50 Hz a.c. it will turn off at the zero crossing points in the waveform so there could be a delay of about 10 ms in the release time.

## 5.6 Hall effect devices

### 5.6.1 *The Hall effect*

If a thin piece of conductor or semiconductor is placed in a magnetic field and a current is passed through it such that the electrical field is at right angles to the magnetic field, as shown in fig. 5.23, then a voltage will appear across the remaining transverse faces of the material. This voltage is called the Hall voltage, $V_0$, and the effect is known as the Hall effect. Although the Hall effect occurs in both conductors and semiconductors it is much more pronounced in semiconductors.

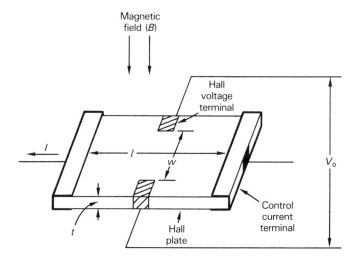

Fig. 5.23. Operation of a Hall effect generator.

The value of the open circuit Hall voltage $V_0$ is given by

$$V_0 = R_H I B / t \qquad (5.13)$$

Therefore the thinner the material the higher the Hall voltage. $R_H$ is called the Hall constant of the material and has units of cubic metres per degree centigrade. It is given by:

$$R_H = 1/n e \qquad (5.14)$$

where $n$ is the concentration of electrons in the material and $e$ is the electron charge.

The Hall effect generator can be considered to have two internal resistive elements, $R_1$ in the electrical control side and $R_2$ in the Hall voltage side. These resistors are discussed later in this section.

### 5.6.2 *Construction of the Hall device*

The effective length $l$ and width $w$ of the Hall plate are important parameters, since (5.13) is only correct for a ratio $l/w$ of infinity. In practice this equation is followed closely for $l/w$ greater than about 2. As an example, if $l/w$ is one i.e. the Hall plate is square, then the Hall voltage is only 50% to 60% of the theoretical value given by (5.13). Generally, there is little to be gained in a practical device in making $l/w$ greater than 3 since there will then be wastage of material.

The material used to make the Hall plate should have several properties:

(i) it should have a high Hall constant and carrier mobility in order to give a high Hall voltage;

(ii) it should have a low electrical control resistance so that large control currents can flow without high power dissipation;

(iii) it must be thin;

(iv) it should have a low temperature coefficient of Hall voltage and material resistivity.

Although both conductors and semiconductors can be used as the Hall plate material, semiconductors have an electron concentration, $n$, of $10^{22}$ m$^{-3}$ which is almost $10^7$ times less than that of conductors. From (5.14) and (5.13) it is seen that semiconductors will therefore give $10^7$ times larger Hall voltages than conductors. The mobility and energy gap of the material are also important considerations. If the mobility is low, then for a given thickness the resistance of the material will be higher, and this will give greater power dissipation and higher temperature rises. This will cause intrinsic carrier conduction and so increase the value of $n$ and reduce $R_H$. The higher the energy gap in the material the higher the temperature at which the Hall plate can work without much reduction in $R_H$.

Materials used in the manufacture of Hall plates include indium arsenide and indium arsenide phosphide. Indium arsenide has a mobility of greater than 33 000 cm$^2$/V s when pure, but it is usually doped to give an acceptable value of temperature coefficient of Hall constant, and this doping gives an $n$ of about $5 \times 10^6$ to $7 \times 10^6$ cm$^{-3}$. The room temperature mobility is now reduced to 20 000 cm$^2$/V s and the Hall constant to 100 cm$^3$/°C. Indium arsenide phosphide has about half the mobility of indium arsenide but a higher energy gap, and it can run at higher temperatures. Indium antimonide is also sometimes used for Hall plates but it has a high temperature coefficient of $R_H$ of about 2%/°C as shown in fig. 5.24.

The Hall plate semiconductor is usually prepared by one of three techniques:

(i) The semiconductor is cut from bars of the material and ground and etched to give wafers of between 5 μm and 100 μm thickness. The semiconductor layer is then stuck onto a substrate by a 1 μm to 2 μm thick layer of

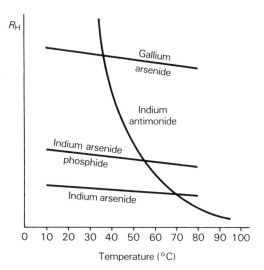

Fig. 5.24. Variation of $R_H$ with temperature.

epoxy resin. The advantage of this is that the adhesive fills the cracks in the material and gives good heat transfer.

(ii) The semiconductor material is vapour deposited onto the substrate. Thicknesses of $2\,\mu m$ to $3\,\mu m$ are attained and the device is suitable for use at very low and very high temperatures.

(iii) Gallium arsenide is grown epitaxially from the gas phase onto semi-insulating gallium arsenide. A layer of about $10\,\mu m$ thickness is used and this has high stability and a low temperature coefficient ($\beta$) so that it is suitable for precision measurement applications.

Compromise is required in the preparation of the Hall plate since most applications require a thin plate (small $t$) but this results in low mechanical strength. The device can be enclosed in sintered ceramic or cast resin for added strength. This usually gives an air gap which needs to be minimised in some applications. An alternative arrangement is to enclose the device in a ferromagnetic jacket; the air gap is then reduced to the approximate thickness of the semiconductor chip. This is called a Ferrite-Hall effect device.

The substrate used for the Hall plate must be rigid in thin sections. It must have a high resistivity, so that it does not short circuit the Hall plate, and it should have a high thermal conductivity to remove heat generated in the Hall plate. The temperature coefficient of expansion of the substrate must match that of the semiconductor used in the Hall plate to prevent stress being set up in the plate due to unequal expansions, and the substrate material must be cheap and easy to shape. Beryllia and alumina are both used as substrates in Hall effect devices.

The electrical contacts can be soldered to the Hall plates if care is taken to prevent curling of the underlying adhesive. A better method is to evaporate the metal onto the semiconductor, as this gives a better control on its dimensions, and then alloy it by heating. The electrical contacts should also extend across the width of the Hall plate so that the lines of current flow are straight and parallel.

Hall electrode terminals should be point contacts although in practice they have finite length. The material used in the Hall contact has a higher conductivity than the semiconductor so that lines of current flow become distorted near the Hall electrodes giving a non-linear relationship between $V_0$ and $B$. The non-linearity can be minimised by special construction techniques, such as by making the ratio of the Hall electrode length to $l$ less than 0.1. Alternatively one can compensate for the non-linearity by loading the Hall generator with a linearising resistor ($R_{LL}$) as described in the next section. This however reduces the sensitivity of the device where sensitivity $S$ is defined as the Hall voltage per unit $B$ and $I$ and is given by

$$S = V_0/I\,B \qquad (5.15)$$

### 5.6.3 *Parameters of Hall effect devices*

The Hall constant $R_H$ is not constant for any device but varies with temperature and with the strength of the magnetic field. Generally, most materials do not show a significant change in $R_H$ for magnetic fields up to about 15 T, except for indium antimonide where the change is large above about 5 T.

The Hall effect device has a control current rating ($I$), stated at 25 °C, which is such that it will give a temperature rise in the semiconductor layer of 10 °C to 15 °C when operated in static air. If the control current is made too large then the device can be destroyed by overheating. The maximum value of the magnetic control field $B$ is stated in data sheets as the range over which there is linearity between the Hall voltage $V_0$ and $B$. The Hall device cannot be destroyed by an excessive magnetic field, although the magnetic field increases the control side internal resistance $R_1$ and this could give higher dissipation. At fields above about 2 T the control current needs to be limited to below its rated value, to keep the semiconductor temperature below 120 °C.

For devices with a ferromagnetic jacket the value of the magnetic flux is given at the value below the saturation level of the jacket material. If the Hall device is used as a modulator or multiplier then the maximum value of the field ampere turns is specified.

The open circuit Hall voltage $V_0$ is given in data sheets at the rated control current $I$ or magnetic flux density $B$. The Hall voltage decreases with load. Data sheets usually give the minimum and maximum values, which can

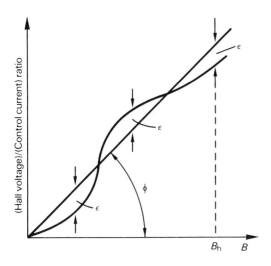

Fig. 5.25. Illustration of linearisation error in a Hall effect device.

differ by up to 50%. Generally, the optimum linearity in the Hall voltage variation with $B$ and $I$ is obtained at a certain value of resistive load, which depends on the Hall generator. This load is called the resistive termination for optimum linearity ($R_{LL}$) and is given in data sheets. Even with this value of load termination the linearity will not be perfect, as shown in fig. 5.25. To define the amount of non-linearity, a line is drawn so that the maximum deviations $\epsilon$ above and below the line are the same. The mean sensitivity $K$ is defined as the slope of this line (tan $\phi$), and the non-linearity $F_L$ is defined as the maximum deviation of the Hall voltage from this line, referred to the end value of the measuring range

$$F_L = \epsilon_{max}/KB_h \qquad (5.16)$$

The control side internal resistance $R_1$ is measured between the control leads when the Hall circuit is open. The data sheets give the resistance value when $B = 0$ ($R_{10}$) and a curve showing how $R_1$ varies with the strength of the magnetic field. Since $R_1$ varies with the magnetic field, and current $I$ determines the Hall voltage, in most Hall effect applications a constant current source should be used to provide the control current, rather than a constant voltage source.

The Hall side internal resistance $R_2$ is defined as the resistance measured between the Hall leads with the control circuit open, and this is also a function of $B$. Data sheets quote the value of this resistance at $B = 0$ ($R_{20}$)

As a result of the process used to make Hall devices, a small resistive component of voltage is usually superimposed onto the Hall voltage. With zero field, $B$, this voltage is given by $R_0 I$ where $R_0$ is called the resistive zero component of the Hall device. The voltage due to $R_0$ can be compensated for in a Hall effect device application by external circuitry.

The lead wires of Hall electrodes form an inductive loop of area $A_0$ which cannot be completely eliminated. If the magnetic flux is changing at the rate $dB/dt$ then even with zero control current $I$ a voltage $E$ will be induced in this loop where

$$E = A_0 \, dB/dt \qquad (5.17)$$

$A_0$ is called the inductive zero component of the Hall effect device and is measured in square centimetres. The value of $E$ varies with $dB/dt$, i.e. the amplitude and frequency of the control magnetic field. For $B = 1$ T and at a frequency of 50 Hz the value of $E$ is typically 500 $\mu$V i.e. $A_0$ is about 0.05 cm$^2$.

Several temperature effects occur in a Hall device. The open circuit voltage, $V_0$, variation is measured as the temperature coefficient $\beta$ and the change in the Hall side internal resistance as the temperature coefficient $\alpha$. The mean values of $\alpha$ and $\beta$ over a range 0 °C to 100 °C are given in data sheets. Generally $\beta$ is used when calculating performance based on no load, and $\alpha$ and $\beta$ are both used for load calculations

$$\alpha \, (\%/°C) = \frac{\Delta R}{RT\Delta T} \times 100 \qquad (5.18)$$

$$\beta \, (\%/°C) = \frac{\Delta V_0}{V_0 T\Delta T} \times 100 \qquad (5.19)$$

### 5.6.4 Hall effect applications

Hall effect devices are used in many applications in industry. Generally these can be grouped into two main areas. In the first, the control current is kept constant and the device is placed in an open magnetic field so that it is primarily used to sample magnetic fields. Since the device is thin it can go into narrow gaps. The Hall voltage can also be used in a feedback system to measure and keep the strength of the magnetic field constant. Applications in this area include magnetic reading heads, and switches with moving magnets.

The second area of application of Hall effect devices uses the component in a closed magnetic circuit. The d.c. current can be variable and the Hall device is used to measure this current. The Hall generator can also be used as a multiplier when two currents are used, one to vary the magnetic field strength and the other to act as the Hall control current. In a modulator application the Hall device is placed in a magnetic circuit in which the field coil is excited by an a.c. signal of a high frequency $f$, and the signal is fed to the control current circuit. The Hall voltage is now alternating of frequency $f$ but modulated by the control current.

### 5.7 Magneto resistors

#### 5.7.1 *The magneto resistive effect*

The magneto resistive effect is seen as an increase in resistance of an element placed in a magnetic field, as the strength of the field is increased. The resistance varies as the square of the flux density $B$ for low values of $B$, and it varies linearly at higher values of $B$. Magneto resistive elements depend for their operation on the law of electrodynamics which says that the Lorentz forces act on mobile charge carriers in a magnetic field and cause the electrons in space to move from a straight path to an indirect path. This results in a lengthening of the current paths, and a narrowing of their cross section, so that the bulk resistance is increased when the electrons move in a solid.

#### 5.7.2 *Construction of magneto resistors*

The extent of the deflection of electrons in a magnetic field depends on the mobility of the electrons in the material. In conductors, the change in resistance (and the Hall voltage) in a magnetic field is low since there is high electron density and this gives a low electron mobility and a Hall angle less than 0.5 degree.

In a semiconductor the Hall angle and the electron mobility are much larger. A metal has a mobility of about $50\,\text{cm}^2/\text{V}\,\text{s}$ whereas the semiconductor with the largest mobility is indium antimonide, with a mobility of $78\,000\,\text{cm}^2/\text{V}\,\text{s}$. With a magnetic field of 1 T

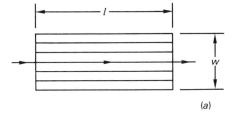

(a)

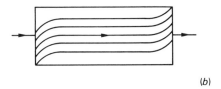

(b)

Fig. 5.26. Current paths in a semiconductor wafer; (a) without magnetic field, (b) with magnetic field.

this gives a Hall angle greater than 80°. Magneto resistors are now mainly made from indium antimonide, and sometimes from indium arsenide which has a mobility of $24\,000\,\text{cm}^2/\text{V}\,\text{s}$.

The dimensions of the magneto resistor has an important relationship to its properties. Fig. 5.26 shows that when a magnetic field is applied to a long thin bar the current path near the centre of the material is similiar to that obtained when no field is applied, but near the edges they are rotated by the Hall angle so that the current path length is increased. The smaller the ratio $l/w$ the less the effect of the central portion of the material, the greater the overall deflection of the current path and the greater the increase in resistance. Fig. 5.27 shows this for several identical specimens of indium antimonide having different geometries.

In order to get a high intrinsic resistance at zero magnetic field, and a large change in resistance with increasing magnetic field, several devices with low $l/w$ ratios can be connected in series. This can be done by depositing a metallic film on a long piece of semiconductor and etching out areas to leave a rastered pattern of short-circuiting conductors (fig. 5.28a). Now the current path is rotated by the Hall angle $\theta$ when a magnetic field is applied (fig. 5.28b). Since the metallic conductors are equipotential surfaces, the current flows back

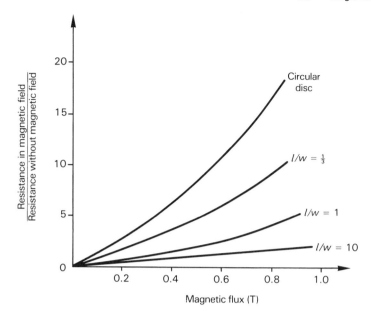

Fig. 5.27. Dependence of magneto resistance on the geometry of the device.

to the other side of the material and reappears rotated by $\theta$ in the next section of the semiconductor (fig. 5.28c). This method of construction allows wide variations in the intrinsic resistance value by a careful choice of the $l/w$ ratio.

Instead of using external metallic short-circuiting lines across the semiconductor the material indium antimonide can be made with built-in parallel oriented crystals of metallic nickel antimonide which act as internal short-circuiting lines. Without a magnetic field the current path is at right angles to the equipotential surfaces of the metallic crystals, as shown in fig. 5.29a. When a magnetic field is applied the current path is rotated by the Hall angle $\theta$ relative to the electric lines of force. As with the rastered system large values of intrinsic resistance can be obtained and this can be varied by controlling the indium antimonide to nickel antimonide ratio, and by doping it with other impurities which vary the material conductivity.

Commercial magneto resistors are built by depositing a film at about 25 μm thickness of the indium antimonide/nickel antimonide onto a substrate of about 0.1 mm thickness. The film is in the form of a meander and the

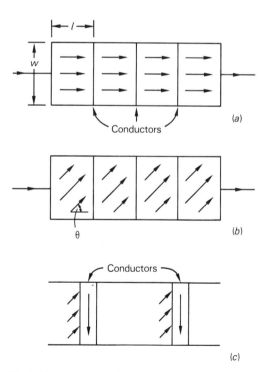

Fig. 5.28. Current path in a rastered magneto resistive arrangement; (a) without magnetic field, (b) with magnetic field, (c) detail of conductor section of (b).

## 106  Magnetic components

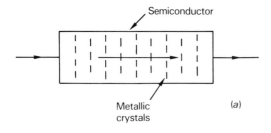

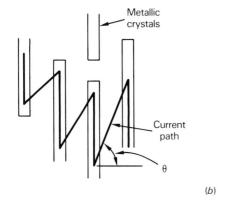

Fig. 5.29. Semiconductor with internal short-circuiting metallic crystals; (a) current flow without magnetic field, (b) detail of current flow with magnetic field.

value of resistance at zero magnetic field can be varied by changing the dimensions of this meander and the number of loops. The film is insulated from the substrate, which may be a magnetic base coated with a thin insulating film or a non-magnetic base such as ceramic or plastic.

### 5.7.3 Magneto resistor parameters

The value of the resistance depends on the direction of the magnetic field, as shown in fig. 5.30. The maximum change in resistance occurs when the nickel antimonide crystals are parallel to each other, and the magnetic and electric fields are at right angles to each other.

The temperature response of the magneto resistor depends very largely on the doping of the material. Generally the resistance decreases with temperature and the drop in resistance is greater after the application of the magnetic field so that the ratio of the resistance in a magnetic field to the resistance without a magnetic field decreases with temperature.

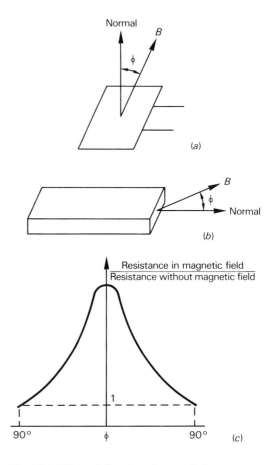

Fig. 5.30. Effect of direction of magnetic field on resistance in a magneto resistor; (a) and (b) field direction in different structures, (c) resistance variation.

The magneto resistor is rated at a given power dissipation, and data sheets give its thermal conductance in free air and when mounted on a heat sink. The device must be derated at higher ambient temperatures to keep the semiconductor temperature from exceeding 150 °C.

Magneto resistors have low noise unless cracks appear in the material due to mechanical strain resulting from faulty mounting. Since the device is a bulk resistor it does not depend on surface effects so that there is very little ageing. Most of the ageing is due to deterioration of the epoxy resins used to pot the device or to stick the semiconductor to the base.

# 6. Peripheral components

### 6.1 Introduction
The components described in this chapter may be looked on as being peripheral to the main design. These components include switches, heat sinks, fuses and the interconnections between boards.

### 6.2 Switches and keyboards

#### 6.2.1 *Switch parameters*
A switch has parameters similar to those of a relay (see section 5.5) except that the operating mechanism is entirely mechanical and there is no electrical energisation. The operating force can be sideways as in slide and lever switches, or in and out as in push button switches. Fig.6.1 illustrates some of the terminology used in switches. A push button switch is shown in four positions during its cycle. With no force applied the switch rests in its free position and the switch contacts are not operated. As the force increases the plunger reaches the operate position at which the switch contacts open or close, depending on the type of switch. Further operating force will move the plunger to its total travelled position, where it cannot be further depressed. Any more force could damage the switch mechanism. When the operating force is reduced the plunger returns to the free position through the release position at which the switch contacts go back to their original position.

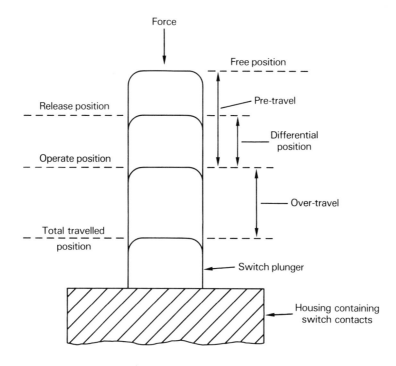

Fig. 6.1. Operating terminology used for a switch.

The switch operating force, or actuating force, is the minimum force needed to close the switch. The release force is the value to which the operating force must be reduced to open the switch. The switch differential force is the difference between the actuation and release forces. For switches which work with a rotary action it is more usual to specify the operating and release forces in terms of torques. In both cases these forces will change with the life of the switch mainly due to mechanical wear.

All switches have sufficient lubricant sealed into them to last their expected life. Operation at high temperatures reduces the viscosity of the lubricant which may cause lubricant loss, so that damage occurs to the mechanical parts.

Current switching and current carrying characteristics of switches are similar to those of relays. The switch can carry a larger current than it can switch due to arcing of the contacts, and the contact resistance decreases with contact current so that the ability to operate reliably on low contact currents is as important a consideration as maximum ratings. Contact bounce contributes to contact arcing and it increases with the size of the moving member and with its speed of operation.

The contacts used for switches must have a large surface to break or carry the load current, and a small mass to reduce contact bounce. They need to have a separation in the open position sufficient to withstand the rated voltage, a high pressure when closed to reduce contact resistance and bounce, and to give good resistance to shock and vibration. The contact material should have good electrical and thermal conductivity, have resistance to electrical and mechanical wear, be free from surface films, and have low cost. Silver and silver based materials are generally used since these have high thermal conductance and the highest electrical conductance. Gold coating is used for dry circuit applications since it is resistant to the formation of oxide and sulphide films. Usually, the gold is separated from the silver by a barrier layer of dense metal like nickel which prevents the gold from migrating into the silver.

The electrical life of a switch is very dependent on the load current and power factor. When current is switched an arc is formed which melts and vapourises part of the contacts. At low atmospheric pressures, caused by higher altitudes, the air ionises at a lower voltage than at sea level and this keeps the arc going for a longer time so that the switch current and voltage ratings are reduced. The metal vapour causes loss of contact volume and can eventually lead to switch failure. The vapour also settles on the housing and this causes increased leakage along the surface of the switch between the contacts, which can lead to overheating and insulation breakdown. Therefore, careful selection of switch housing (insulation) material is required for different applications.

Switch failures manifest themselves in three forms, as catastrophic failures, as intermittent failures and as parametric failures.

A catastrophic failure may be due to: (i) collapse of the operating mechanism caused by a looseness or tightness of the actuating system; (ii) voltage breakdown caused by overheating or metallic deposit of contact material on the insulating surface; (iii) contacts becoming welded together, caused by excessive arcing.

Intermittent failures may be due to: (i) a fault in the switch actuator caused by incorrect mounting or excessive wear of the mechanism; (ii) intermittent high contact resistance caused by excessive arcing or low contact force; (iii) intermittent delayed contact action due to excessive arcing or metal transfer between contacts.

Parametric failures may be due to: (i) excessive operating force caused by wear or distortion of the switch mechanism; (ii) high contact resistance caused by excessive arcing or metal transfer; (iii) reduced insulation resistance caused by overheating, moisture absorption, or surface contaminations including metal deposits.

### 6.2.2 Construction of switches

There are many different types of switches but generally they can be grouped into two classes, those which are mounted externally on an equipment and those which are mounted internally.

The external switches will be operated many times during the life of the equipment. They must look attractive when mounted on the front panel of the equipment and they must be easy to identify and operate without interference from other items on the panel.

Switches used internal to the equipment are usually for adjustment purposes and will usually be used by skilled operators, and operated only a few times during the life of the equipment.

#### 6.2.2.1 *Positive drive*

Probably the simplest switch mechanism is the positive drive system, three variations of which are shown in fig. 6.2. In figs. 6.2a and c with the operating arm in one direction contact B is closed and when the arm is moved to the other direction B opens and A closes. The rotary switch (fig. 6.2b) has one moving and several fixed contacts but the principle of operation is the same. There is positive drive between the operating arm and the moving contact. This means that the operator controls how fast and by how much the switches open and close.

Positive drive is desirable for safety reasons since even in instances of severe contact welding it should be possible for the operator to apply sufficient force to open the switch. However it has several disadvantages: (i) the switch can be damaged easily by misuse in applying too much force on the contacts; (ii) the contacts can be 'teased' into a half open position which results in excessive arcing; (iii) the contact separation time may be long resulting in excessive arcing; (iv) there is poor repeatability of the make and break positions since contact wear and spring fatigue change the contact operating positions.

The *snap-action switch* overcomes most of the limitations of a positive drive switch.

#### 6.2.2.2 *Snap-action switch*

Fig. 6.3 shows the operation of a snap-action switch. In the free position the moving contact is held against fixed contact A by the action of the spring. Just past the operating position the spring causes the moving contact to change over rapidly from contact A to contact B. There are many versions of the snap-action switch such as toggle, push button, rotary and sliding.

The snap-action switch gives precise control over the operating and release points and the force required to operate the switch. The change-over time is independent of the actuation velocity so that no contact teasing can occur. A disadvantage of the snap-action switch is that there is no positive drive between the operating arm and moving contact so that only the spring force is available to break the contacts if a fault occurs. Another disadvantage is that due to the high operating force of the spring there is considerable contact bounce.

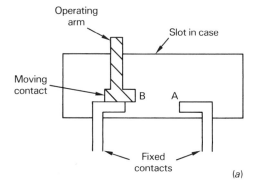

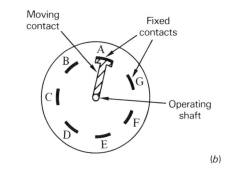

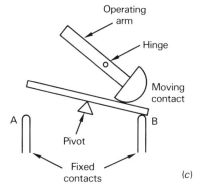

Fig. 6.2. Simple switch arrangements; (a) slide switch, (b) rotary switch (single wafer), (c) rocker switch.

#### 6.2.2.3 *Dual-in-line switches*

There are many types of miniature switches

110  *Peripheral components*

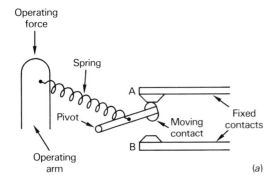

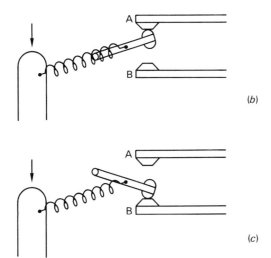

Fig. 6.3. Snap-action mechanism; (a) free position, (b) operating position, (c) total travelled position.

designed for mounting on to printed circuit boards. The dual-in-line (DIL) switch can have 14 to 16 pins and contains an array of single pole single throw switches, or fully coded 10 and 16 position units. The switch mechanism can be rocker, slide, toggle, lever or rotary.

Problems can arise during printed circuit board flow soldering if a switch is mounted on it. Flux can enter the body of the switch and cause malfunctions. This is avoided by hand soldering the switch or by using a DIL socket and inserting the switch at a later stage. Some designs use sealed switches but if contamination gets into these it cannot escape. Generally, DIL switches have a relatively low operating temperature range.

Since a DIL switch normally operates infrequently during its life it can malfunction due to oxide and insulating films forming over its contacts. Therefore, the contact material needs to be carefully selected, especially when the switch will be operating at low currents and voltages. A good switch uses a 1 to 3 $\mu$m layer of gold over a 2 to 3 $\mu$m nickel barrier layer, and the contacts operate with a wiping action.

### 6.2.2.4 *Rotary switches*

The simple switch shown in fig. 6.2b can be expanded by connecting several wafers to one shaft to give a multitude of poles and switch combinations. The switch can also be coded as shown in fig. 6.4 where a row of four brushes is used to move over the mating tracks on the board. The brushes are connected together and form one terminal of the switch. Where a track exists for that brush position then the corresponding edge connector will be connected to the brush to give the binary coded decimal (BCD) code shown in fig. 6.4a. Since the codes are produced by printing onto a printed circuit board a large number of different codes can be obtained at low cost and a code can be changed relatively easily by changing the board.

In an alternative form of rotary switch called a thumbwheel switch the brushes are rotated by attaching them to a wheel and moving the wheel by its edge. It is also possible to keep the brushes stationary and to rotate the printed circuit board, with its printed pattern, past the brushes. The printed tracks are usually electroplated gold for low level switching, with a nickel plating under the gold. In some arrangements a lever is provided instead of a wheel, and this makes operation in certain circumstances, for example with a gloved hand, much easier since the lever can be grasped and turned through a 90° arc.

Thumbwheel switches require about half the amount of operating torque required by shaft type rotary switches. They are easy to read and present a one digit display at a time. Many types of output connections are available, such as solder tags and printed circuit board edge connectors. There are also facilities for mounting additional components, such as diodes, on to the back of the printed circuit board.

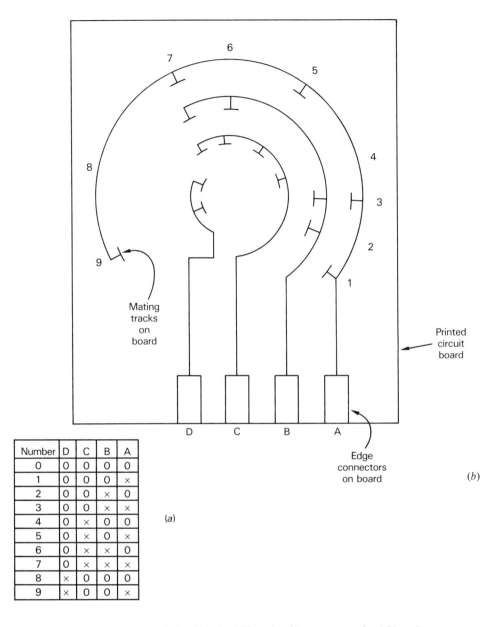

Fig. 6.4. Rotary coded switch; (a) BCD code, (b) arrangement for BCD code.

### 6.2.2.5 Limit switches

A *limit switch* is designed to sense the position of a mechanism and to operate a contact when a set position is reached. Fail-safe operation is often needed so there must be positive drive between the operating force and the moving contact.

Fig. 6.5a shows one type of limit switch using a long actuating arm. The plunger operates the switch and the use of the arm means that the actuating force has been multiplied by the factor $l_1/l_2$. The limit on $l_1$ is reached when the switch can operate under the weight of the arm. Apart from amplifying the actuating force the actuating arm allows the switch to be mounted away from the direct line of the force, and the force moves through a larger distance than the plunger, as may be required in some applications.

Fig. 6.5b shows an alternative arrangement which is mainly used for cam operation. The roller reduces wear and if a nylon roller, which is lighter than a steel roller, is used, then the actuating arm can be made long to give amplified force. Both bidirectional and unidirectional switches are available. A unidirectional switch is shown in fig. 6.5b in which the roller carriage is hinged so that it lies back when the force is applied in the non-actuation direction. The limit switch may be made even safer by using insulating material to make the actuation arm.

The ideal limit switch is that which combines a snap-action section to give rapid make and break of the load circuit, plus a positive drive section for back-up protection.

### 6.2.2.6 Proximity switches

A *proximity switch* is similar to a limit switch

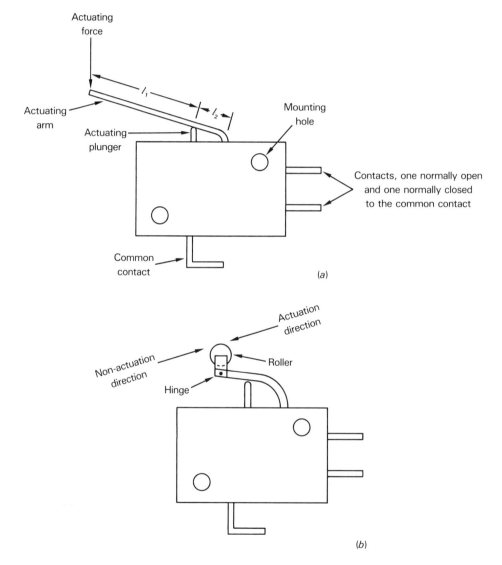

Fig. 6.5. Mechanical limit switches; (a) lever type, (b) roller type.

in that it senses the position of an object. However, in a proximity switch there is no mechanical contact between the switch and the object being sensed. There are two types of proximity switches, magnetic and inductive.

Magnetic proximity switches can be made from reed relays and magnets. A bias magnet may be used to keep the reed closed in the absence of an actuation magnet if a normally closed switch is needed. A ferrous metal can also be sensed since when the ferrous metal approaches the magnet it diverts some of the flux away from the reed, which then operates. This switch has hysteresis since the ferrous metal will acquire some magnetism and will therefore have to move further away to cause the reed to release. The effects of hysteresis can be reduced by special mounting techniques.

An inductive proximity switch uses an oscillator, a detector and a switch. The oscillator generates an oscillating magnetic field which induces eddy currents in any electrically conducting material which moves near to it. This causes the oscillations to be damped, and the detector senses this and switches the output switch. The range of the proximity switch can be easily adjusted by changing the $Q$ value of the circuit, a lower $Q$ value giving a lower distance over which the switch can sense an object. However this also makes the device sensitive to objects approaching it from the sides.

The magnetic proximity switch has a shorter life than an inductive switch, and can only detect ferrous metals or magnets. The sensing field can permeate various dielectric materials such as glass, plastics and wood. An inductive proximity switch has a much longer life and can operate faster than a magnetic switch with very little hysteresis, and good control over the sensing range. It is more expensive than a magnetic switch and has a tendency to fail into the 'on' position since a power failure will stop oscillations and this can be mistakenly taken to indicate the presence of an object which has damped the oscillations.

### 6.2.2.7 Keyswitches

The switches described in this section are those used with keyboards. There are many different types of keys but generally they work on a contacting or a non-contacting principle. In a contacting key the current is mechanically switched, whereas it is electrically modulated in a non-contacting key such as a Hall effect device. An example of a contacting key is the reed keyswitch shown in fig. 6.6. The side magnet arrangement gives a switch of low profile. Interaction between adjacent switches is avoided by using a magnet which is physically short compared to the reed. In the rest position the magnet is in contact with the soft iron keeper and this ensures small differential between the on and off positions and good repeatability of the switch. The end magnet

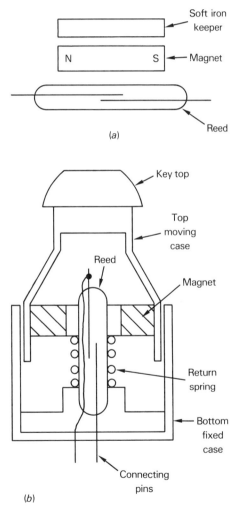

Fig. 6.6. Reed keyswitch; (a) side magnet, (b) end magnet.

114  *Peripheral components*

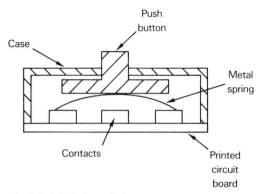

Fig. 6.7. Bubble keyswitch.

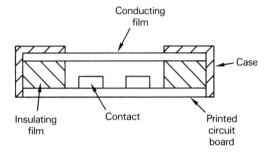

Fig. 6.8. Conductive film keyswitch.

arrangement practically eliminates interaction with adjacent switches, and the effect of stray external fields.

The reed keyswitch is sealed and has a good operating life but suffers from high contact bounce. This bounce can be reduced, and the operation of the switch on dry circuits improved, by using mercury-wetted types of switches.

Fig. 6.7 shows a mechanical keyswitch incorporating a bubble shaped metal spring. The push button rests on this spring and when it is depressed the spring bends and touches the contacts on the printed circuit board so that the switch is closed. The spring is usually made from phosphor bronze or beryllium copper. This switch arrangement is shallow and it has a positive action in its operation.

Switches made from conductive rubber or plastic films have low bounce and long life. Fig. 6.8 shows such an arrangement and, like the switch of fig. 6.7, many switches can be combined onto a printed circuit board to form a low cost keyboard. The switch is operated by depressing the conducting film until it touches the contacts on the board. There is very little film movement during the depression and no positive action, the switch having a very spongy touch. The switch is very shallow having a depth of 5 mm to 10 mm.

The conductive film can be made from carbon elastomer or silver elastomer. Carbon elastomer has a higher contact resistance and gives less sharp switch operation, as shown in fig. 6.9, but it is cheaper than silver elastomer. Since silver has a tendency to migrate it is usually applied to resistant material such as urethane. Fig. 6.10 compares the properties of the two types of materials.

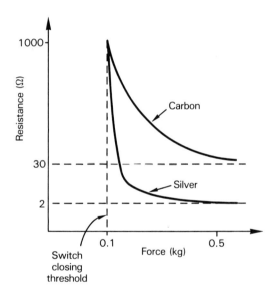

Fig. 6.9. Switching curves of conductive film switches.

| Parameter | Carbon | Silver |
|---|---|---|
| Base material | Silicone | Urethane |
| Contact resistance ($\Omega$) | 30 | 2 |
| Maximum current (mA) | 25 | 75 |
| Contact bounce (ms) | 1 | 0.5 |
| Current rise and fall times (ms) | 1 | 0.1 |

Fig. 6.10. Typical characteristics of carbon and silver elastomer films.

*6.2 Switches and keyboards* 115

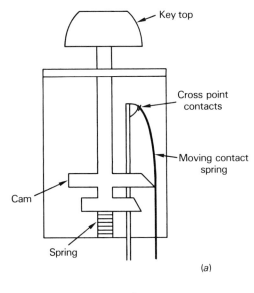

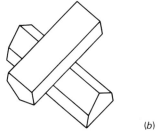

Fig. 6.11. Cross point switch; (*a*) switch construction, (*b*) cross point contact arrangement.

Fig. 6.11 shows another mechanical contacting switch arrangement. Depressing the key causes the cam to be forced against the moving contact spring and to open the switch. The cross point contact arrangement shown in fig. 6.11*b* is such that the contact area is very small, so that contact power is increased and contact resistance is low.

Non-contacting keys operate mainly on magnetic core, Hall effect, optical or capacitive principles. The Hall effect key consists of a magnet attached to the key top. When the key is depressed the magnet moves past an integrated circuit containing the Hall effect device. This produces an analogue voltage in the device, which is amplified and a pulse of about 1 ms duration is generated by the switching electronic circuitry. The Hall switch, like many other non-contacting switches, has no mechanical contacts,

no bounce, long life and high reliability. However it is more expensive than the cheaper contacting switches, and needs power to operate the switching electronics.

The optical switch is made in many forms but basically it consists of an optical source and detector with a moving shutter between them. Depressing the key top will either move the shutter between the source and detector, or move it away.

A magnetic core switch consists of a magnet, connected to the key top, which can move past a tiny ferrite core through which two single turn coils are wound. An a.c. signal is applied to one of the coils and this is sensed on the other coil. When the key is not pressed the magnet is near the core and it is saturated so that the a.c. signal is not coupled from one coil to the other. When the magnet is removed by pressing the key the core is no longer saturated so that the a.c. signal is coupled through and can be rectified and fed to associated sensing circuits.

A capacitive key consists of an upper disc which is deflected to increase the capacitive coupling which occurs between two lower contacts, without touching them. This increases the capacitive coupling between the contacts such that a signal source on one contact is fed through to the other. The capacitance of the switch varies from about 5 pF in the unoperated state to about 10 pF in the operated state and this change takes place in about 0.1 ms. The disadvantage of this switch is that, like the contacting switch, the switching action takes place near the end of a key stroke.

### 6.2.2.8 *Keyboards*
Keyboards consist of a large number of key switches and associated electronics which produce signals for the keys and scan the keys every 2 to 3 ms to detect when they operate. Usually the keyboard produces a coded output such as binary coded decimal (BCD) which uses four bits, European standard code for information exchange (ECMA) which uses seven bits, and the American standard code for information interchange (ASCII) which uses seven bits. A seven bit output code needs five contacts per key or about 350 contacts for a typical alphanumeric keyboard.

Keyboards operate in mono-mode when a single code is assigned to each key. When the keyboard is in multi-mode more than one symbol is assigned to a key and the function of the key is changed by using shift keys and control keys. Multi-mode functions save panel space since fewer keys are needed but reduce throughput since the operator needs to press an extra key when shifting modes.

A feature often found in keyboards is that of *N-key rollover* and *N-key lockout*. Its operation is shown in fig. 6.12. The keys are pressed in the sequence shown in fig. 6.12a and they produce overlapping keyboard input signals as shown in fig. 6.12b. An *N*-key lockout system would prevent any keyboard output signals when two or more keys are depressed simultaneously as this could indicate an error. This is shown in fig. 6.12c. However, during speed typing the operator would often depress keys in sequence and the *N*-key rollover facility permits this providing there is a specified minimum delay between key depressions. It works as shown in fig. 6.12d by generating a pulse for a duration $T$ at the start of each key depression.

There are several human engineering aspects of keyboard layout. The keytop size should be about 1.25 cm in diameter with a centre to centre spacing of 2 cm. The keyboard force varies from 20 to 200 g and the key displacement from 0.05 cm to 0.5 cm. Introducing switch feedback to show when it operates or increasing keyboard size has little effect on the performance of the operator, unless the keyboard is made too small which increases errors. The slope of the keyboard also does not have much effect on throughput although a 10° to 35° slope is usually preferred.

### 6.3 Fuses

#### 6.3.1 *Operation and construction*

A fuse is used to protect a circuit against current overloads. The fuse is placed between the circuit and the current source, and will open and isolate the circuit from the source if the current through it becomes greater than its rated value.

The fuse consists of a metal element which carries the normal steady state load current, but overheats and melts if a fault current flows which is large enough and lasts for a sufficiently long time. If the current is $I$, the resistance of the fuse element $R$ and the fuse melts after time $t$ then the energy needed to blow the fuse is $I^2 R t$. Since the resistance of the fuse is indeterminate it is usual to refer to the $I^2 t$ rating of the fuse.

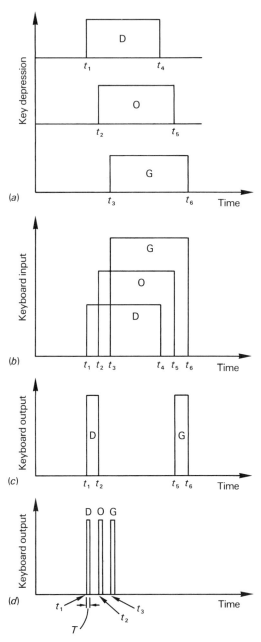

Fig. 6.12. Illustration of the *N*-key rollover principle; (*a*) key depression sequence, (*b*) keyboard electronic input, (*c*) keyboard output with *N*-key lockout, (*d*) keyboard output with *N*-key rollover.

Fig. 6.13 shows the construction of a high rupturing capacity (HRC) fuse. The fuse element is made from pure silver and usually has the shape shown in fig. 6.14. The V-notch structure is preferred to other shapes since it gives a fuse with a greater current capability, whilst having a reduced $I^2 t$ rating so that it will melt quickly if a high fault current occurs. The characteristics of the fuse are determined by the shape and dimensions of its element. A small alloy pellet is sometimes attached to the element. This forms a eutectic alloy with the silver of lower melting point than the silver, so that this part of the element will melt when an overload occurs. A large short-circuit current flow will still cause $I^2 t$ heating at the constrictions of the fuse and these will melt first. In a low current fuse it is difficult to make the constrictions in the element narrow enough so plain wire is often used.

The body material used for the fuse must have good mechanical strength and be able to withstand thermal shock and the high temperatures which arise during normal running. Low

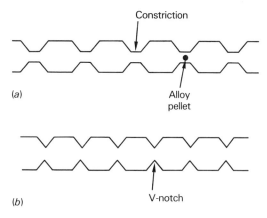

Fig. 6.14. High rupturing capacity (HRC) fuse elements; (a) normal industrial fuse, (b) semiconductor protection fuse.

current (below 10 A) and low voltage (below 200 V) fuses can use glass, but for high power levels ceramic, or sometimes silicon bonded glass fibre, is required to prevent the case from shattering during fuse rupture. A cordierite and steatite ceramic combination is used for general purpose applications where a large fuse size is acceptable. For semiconductor protection applications the fuse size for a given rating must be small and ceramics with a high alumina content are preferred as they have increased strength.

The fuse body is often filled with quartz granules which give rapid heat conduction from the fuse element to the case. The filler also extinguishes the arc by fusing with the resultant silver vapour to form a non-conductive material called fulgurite, and distributes the high pressures generated in the fuse link during overload conditions over the ceramic body.

The end cap and end terminals must give good electrical contact and form a good fit with the inner cap. The material used is usually brass with a high copper content. The end terminal is welded, soldered, or riveted and soldered to the end cap. The construction used gives good electrical contact and isolates the element from external mechanical shocks and vibration.

### 6.3.2 Fuse characteristics

Fig. 6.15 shows the circuit waveforms under conditions of short circuit or heavy current overload. $I_1$ is the peak fuse let through current

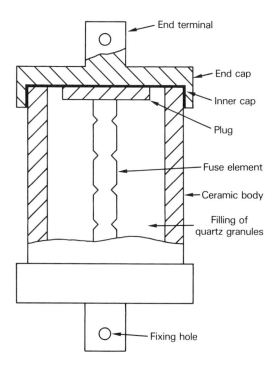

Fig. 6.13. Construction of a high rupturing capacity (HRC) fuse.

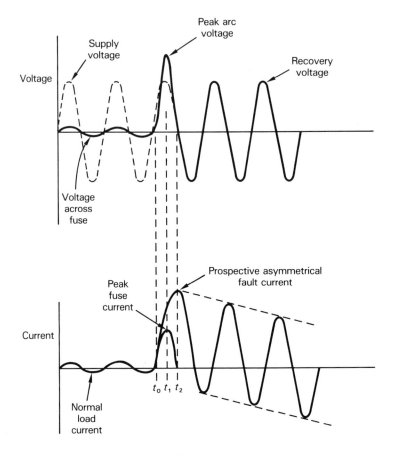

Fig. 6.15. Typical short-circuit voltage and current waveforms.

during rupture. The temperature rises rapidly at the constrictions of the fuse element, and at each constriction an arc is formed and the metal vapourises. The arc voltage is greater than the supply voltage. Time $t_0$ to $t_1$ is called the pre-arcing time and $t_0$ to $t_2$ is called the total clearing time.

The higher the prospective fault current the shorter the pre-arcing time becomes and the higher the peak current before fusing. The peak let-through current is also dependent on the fuse arc voltage and the system voltage.

For a fuse used to protect a semiconductor device the $I^2 t$ rating of the fuse must be less than that of the semiconductor being protected. The $I^2 t$ rating varies with the magnitude of the fault current, and with the operating voltage. The arcing voltage is usually designed to be low, by shaping the constrictions in the fuse element so that the arc is extinguished relatively slowly, but then the $I^2 t$ rating is adversely affected.

A fuse should be chosen to have a lower rupturing time characteristic than the device being protected. For any value of current overload the fuse must rupture in a shorter time than the device being protected. The fuse is rated in rms current since this determines the heating effect. The circuit waveforms may not always be sinusoidal, for example a half-wave rectified or chopped wave may be used, so that one can have a high ratio of rms to average current. The fuse should obviously be chosen such that it does not rupture when the circuit is working within its average rated current.

The voltage of the fuse at rupture must be low to avoid destroying adjacent components. Careful choice of the constrictions in the fuse element, and of the packing density of the quartz granules shown in fig. 6.13, is necessary so that no rapid change of current occurs during arcing. There is therefore a compromise between the energy let through during the arc and the voltage caused by rapid current cut-off. The fuse must also be able to clear the expected peak voltage or there will be a continuous arc if the supply voltage is too high.

The semiconductor fuse is required to protect components which have very similar characteristics to itself, such as a low thermal mass, and a small margin between the operating and destruction temperatures. This means that careful design is needed to choose the correct fuse. The semiconductor fuse operates faster and at a lower arc voltage level than a standard fuse.

The rms fuse rating is usually given at 25 °C ambient and needs to be derated at higher temperatures. The end caps should be kept as cool as possible by mounting on a large bus bar or heat sink.

The minimum fusing current is defined as the value of current which will cause the fuse element to melt within four hours. The fusing factor is the ratio of the minimum fusing current to the rated current and it defines the overload capacity of the fuse. A typical value of fusing factor for an industrial fuse is 1.5. If the fuse is used to protect against overloads rather than, as is more common, against short circuits then the fusing factor is important and in some types of loads which cannot stand much overload e.g. PVC wire, the fusing factor must be less than 1.5.

The performance of a fuse can be improved by connecting several of them in parallel. The total steady state current is given by:

Total current = Current of one fuse × N × F

where $N$ is the number of fuses in parallel and $F$ is a factor which accounts for fuse mismatch and is typically 0.9. The $I^2 t$ rating of the combination is given by:

Total $I^2 t$ rating = $I^2 t$ of one fuse × $N^2$

Therefore if two fuses are connected in parallel each is required to have about half the steady state rating of one fuse but the $I^2 t$ rating is improved by a factor of four. Fuses can be connected in parallel by having several fuse elements in the case shown in fig. 6.13, the elements being connected to the same end caps.

## 6.4 Heat sinks

### 6.4.1 *The principles of cooling*

A semiconductor device dissipates heat, and unless this heat is removed the semiconductor temperature can rise above its maximum permissible value and it will be destroyed. Fig. 6.16 shows a semiconductor placed next to a heat dissipator. This dissipator is usually called a heat sink since the heat is absorbed or 'sunk' into it before being dissipated into the adjoining environment.

The efficiency of a heat sink is increased as its area ($A$) increases and its distance ($l$) from the semiconductor decreases. If any separators are required between the semiconductor and the heat sink then these should be placed at the bottom of the pyramid, shown in fig. 6.16, rather than the top since it will then have a greater heat dissipating area.

Thermal circuits can be analysed by analogy with electrical circuits. If $Q$ is the thermal power in watts being dissipated by the device and $\Delta T$ is the temperature difference across the device in degrees centigrade then the

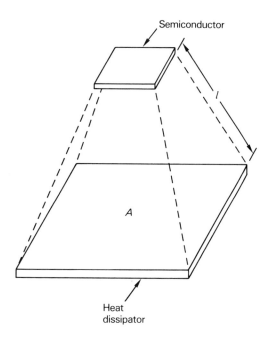

Fig. 6.16. The semiconductor–heat sink pyramid.

*thermal resistance* of the device is given by $\theta$ where

$$\theta = \Delta T/Q \quad (°C/W) \tag{6.1}$$

A complex thermal assembly, such as an encapsulated semiconductor mounted on a heat sink, can be broken into its separate parts and then analysed using (6.1). Fig. 6.17 shows the equivalent circuit of such an assembly where $\theta_{JC}$ is the thermal impedance between the semiconductor junction and the semiconductor case, and $(T_J - T_C)$ is the temperature difference between junction and case. $C_{JA}$ is the thermal capacitance of the junction to the ambient. If the power flowing between junction and case is $Q$ then, from (6.1)

$$\theta_{JC} = (T_J - T_C)/Q$$

Similarly, the other impedances from case (C) to heat sink (S) and heat sink to ambient (A) can be analysed. Generally the thermal capacitances can be ignored in rms calculations and are only used for transient analysis. The thermal resistance between case and ambient $\theta_{CA}$ is usually large compared to that through the heat sink and it can be ignored. The equivalent circuit therefore simplifies to three thermal resistances in series. For this total system

$$\theta_T = (T_J - T_A)/Q \tag{6.2}$$

or

$$\theta_T = \theta_{JC} + \theta_{CS} + \theta_{SA} \tag{6.3}$$

The heat sink loses its heat by conduction to adjoining parts, by convection and by radition. The heat sink can be cooled by natural convection, by forced air which is blown over the heat sink from a fan, or by immersing the heat sink in a liquid coolant.

Thermal analysis can give up to 25% errors in many cases. This is due to: (i) the mix of heat transfer modes and the difficulty of predicting the actual heat transfer path; (ii) the variation in power dissipation between semiconductors of the same type, even when these come from the same batch (power dissipation will also vary due to differences in clamping pressures, and these are difficult to forecast); (iii) many of the constants used in thermal equations are actually low-order variables and choosing the right value of these variables often needs judgement and experience.

Analysis of forced air cooled systems gives less accurate results than analysis of natural air cooled systems because: (i) there are differences in flow over interior and exterior surfaces; (ii) it is not possible to calculate the air velocity at each point in the flow path; (iii) thermal analysis usually assumes symmetrical shapes, e.g. cylinders and spheres, and in practice these shapes rarely occur.

When power is applied to a cold heat sink its thermal resistance is initially very low and it reaches its steady state value after a certain time, which can be several hours for a large heat sink. This gives rise to the concept of a transient thermal resistance, which is used for calculations if short time values are needed. The steady state thermal resistance of the heat sink decreases as the power dissipation increases, and it also decreases as the air flow rate increases in a forced air cooled system.

### 6.4.2 *Construction of heat sinks*

The heat sink is usually made from aluminium and is designed such that there is a large surface area for radiation and convection of heat, and the weight is minimised. The heat sink may be left bright, but coloured matt surfaces are more

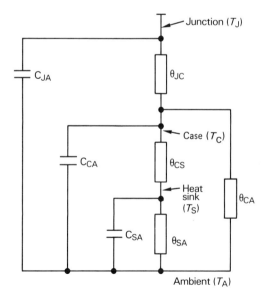

Fig. 6.17. Equivalent circuit of a semiconductor device mounted on a heat sink.

efficient. Black is not necessarily the best colour since at the temperatures being considered heat radiation occurs in the infrared region and all enamels, varnishes, anodised surfaces and oil paints have high emissivities regardless of colour.

Heat sinks are usually designed with fins. The greater the number of fins the larger the area for convection cooling, but if the fins are too close together there is less heat radiation. Therefore a compromise is needed in the heat sink design. Forced air cooled heat sinks are less dependent on fin spacing than natural air cooled ones. The air flow should be designed to create turbulence over the surface of the fins and break up any layer of static air.

Heat sinks may generally be considered to be of three types:

(i) Those used with small electronic components mounted on printed wiring boards. These heat sinks are usually made from sheet metal and their size is critical since space must be allowed around them on a board. The board must also be oriented to give the best air flow over the heat sinks.

(ii) Those used with medium power components such as power transistors. The heat sink is usually made from aluminium extrusion and is designed for a class of components and then machined to take a particular component from that class. The heat sink is sometimes required to provide physical support for the device being cooled.

(iii) Those used with high power components such as power thyristors and diodes. Very stringent cooling needs are placed on these heat sinks and they must be rugged, and are often physically large and expensive.

Electrical isolation is often needed when a device is mounted to a heat sink, and this can be obtained by using isolating washers. Beryllia is the most expensive material used for these washers but it has the highest thermal conductivity and dielectric strength. Hard anodised aluminium is the second best for thermal conductivity and dielectric strength. Mica washers have been used for a long time but they suffer from the fact that they can crack and peel, and because they are transparent it is

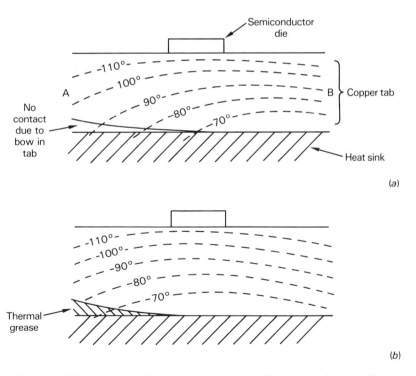

Fig. 6.18. The advantage of thermal grease; (a) temperature contour without thermal grease, (b) temperature contour with thermal grease.

easy for two to become stuck together and used together, which would cause an increase in the thermal resistance and a possible device failure. High temperature plastic washers, such as Kapton and Mylar, are replacing mica. They have half the dielectric strength but they are cheaper and since they are coloured their shading gives a visual indication if two are stuck together. An alternative to using an isolating washer is to spray the heat sink during manufacture with an electrical insulation material

The interface between the case of the component being cooled and the heat sink has a relatively low thermal resistance compared to other parts of the system. However the resistance can increase during assembly by a factor of 10 times unless care is taken to minimise it. This is done by keeping the mating surfaces clean and by applying adequate mating pressure, and by using a thermal grease between them. This grease, or heat sink compound, is a silicone material filled with heat conductive metal oxides. The grease must not dry out, melt or harden even after operating for long periods at high temperatures such as 200 °C. Fig. 6.18 illustrates how the grease helps to even out the temperatures in the semiconductor package. Without the grease a slight bump, bow or dust particle causes the temperature at A to be higher than that at B. The thermal grease replaces the air and has a much lower thermal resistance so that the temperatures in the copper tab, and therefore the semiconductor, are more even. Fig. 6.19 compares the thermal resistance of some commonly used components.

| Material | Thermal resistance (°C cm/W) |
|---|---|
| Diamond | 0.02 to 0.1 |
| Copper | 0.3 |
| Aluminium | 0.5 |
| Solder | 2.0 |
| Thermal grease | 130 |
| Mica | 150 |
| Mylar | 400 |
| Still air | 3000 |

Fig. 6.19. Comparison of thermal resistances of some typical materials.

Diamond is used to cool some specialised components operating at high frequencies.

### 6.4.3 *Heat pipes*

A heat pipe conducts heat from the source to another region where it can be dissipated more easily. It is used to remove heat from inaccessible positions or to take it to a larger remote heat dissipator.

A metal bar conducts heat very inefficiently. For example conducting 1 kW of heat in a solid copper rod of 1½ cm diameter over a 30 cm length would give about 800 °C difference between its ends. A heat pipe of the same dimensions would give a 2 °C difference, therefore it is much more efficient.

The first heat pipe was patented in 1942 but it did not find widespread use until it was adopted for space applications. It is now commonly used in many industrial areas. Fig. 6.20 shows the construction of a heat pipe. A hollow metal tube is sealed at both ends and its walls are lined with a wick material. The inside of the tube contains a small quantity of a working fluid which is in a partial vacuum so that it boils at a lower temperature than it would at atmospheric pressure. The component to be cooled is attached to the evaporator end of the tube. The working fluid vaporises and heat is

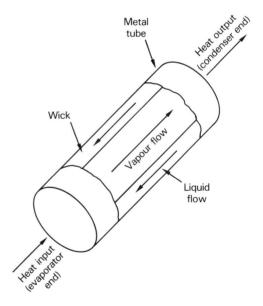

Fig. 6.20. Construction of a heat pipe.

absorbed in converting the liquid to vapour. The vapour travels towards the condenser end of the tube and this end is cooled externally by a heat sink. The vapour gives up its heat at this end as latent heat and condenses. The condensed fluid is returned along the wick by capillary action to the evaporator end. When vapour condenses it tends to increase the vacuum so that more vapour is drawn from the evaporator end. A heat pipe is typically about 0.3 cm to 1 cm in diameter and up to 50 cm long. It can have a variety of shapes to suit the equipment layout and operates in the temperature range +20 °C to +200 °C, although the range −200 °C to +600 °C can be covered if necessary. The power rating of the tube varies from 20 W to 200 W when operated in a horizontal position. A pipe handling 100 W with a working fluid at 100 °C would have a temperature difference between its ends of about 3 °C, and this difference will increase with increasing temperature and decrease with increasing power rating.

The heat transport capability of the heat pipe is very largely determined by the physical properties of the working fluid, such as surface tension, liquid viscosity, liquid density and vapourisation heat. These properties are temperature dependent so the performance of the heat pipe with a given fluid is temperature dependent, and the fluid must be chosen to suit the operating temperature range. The working fluid must also be compatible with the pipe and wick materials. Chemical passivation of the wick and container is required to avoid chemical corrosion. Examples of working fluids (and their working temperatures) are: liquid ammonia (−70 °C to +60 °C); methanol (−45 °C to +120 °C); and water (+5 °C to +230 °C). For electronic components water is the most desirable fluid over the temperature range, and since it is compatible with copper and its alloys the pipe is usually made of copper. Fig. 6.21 shows how the characteristics of some of these fluids change with temperature.

The heat pipe depends for its operation on capillary action to give a pressure difference along its length, which returns the working fluid to the evaporator. Capillary action is destroyed if boiling occurs in the wick. Heat transfer occurs through a liquid filled wick to the container wall, so that the fluid in contact

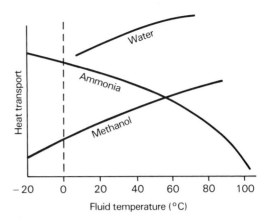

Fig. 6.21. Characteristics of some commonly used heat pipe fluids.

with the inner wall of the container, where the vapour is, is superheated above its boiling point. If this superheating is excessive then vapour bubbles fill the wick and capillary action is stopped. The point at which this occurs depends on many factors, such as wick structure and the inclination of the pipe. The wick also distributes the working fluid over the evaporator surface and so avoids any hot spots.

The power rating of the heat pipe is position sensitive, and it is usually given in data sheets for the pipe in a horizontal position. The rating of the pipe can be almost doubled when its condenser end is vertically above the heated end since gravity then helps the return of the working fluid. The heat pipe capability falls to zero if the evaporator end is raised by more than a value called the static wicking height. This height depends on the working fluid, the temperature and the tube bore size. It is about 15 cm to 40 cm for a typical water filled pipe, the cheaper woven wire mesh wicks giving heads of 5 to 15 cm.

Two equations are used in the design of a heat pipe. The maximum heat transport capability $Q_M$ of the pipe is given by

$$Q_M = Kd^2/l \qquad (6.4)$$

where $d$ is the inside diameter of the pipe, $l$ is the length from condenser to evaporator, and $K$ is a constant which is dependent on the heat pipe geometry, the wick material, the working fluid and the orientation of the pipe. The maximum temperature difference between

## 124 Peripheral components

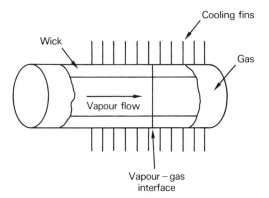

Fig. 6.22. Variable conductance heat pipe.

evaporator and condenser for a power of $Q$ is given by

$$\Delta T = \frac{Q}{b}\left[\frac{1}{A_E} + \frac{1}{A_C}\right] \qquad (6.5)$$

where $A_E$ is the evaporator area, $A_C$ is the condenser area and $b$ is a constant dependent on the configuration of the pipe. Equation (6.5) is used to obtain an approximate design for a heat pipe which then must be tested and modified.

Fig. 6.22 shows a modification to a conventional heat pipe which is used in applications where the source is to be kept at a constant temperature. It is called a variable conductance pipe and contains gas sealed into one end and cooling fins along part of its length. The fins can cool the vapour much more effectively than the gas as the gas is not condensable. The position of the interface between the vapour and the gas will vary with the amount of heat. Therefore, the greater the heat the higher the vapour pressure, which results in a larger cooling surface so that the temperature is maintained substantially constant.

## 6.5 Connectors

### 6.5.1 *Connector parameters*

Connectors vary in size from single point connections to a block having many dozen connections. Fig. 6.23 shows one such system in which electrical contact is made when the pins are inserted into the sockets. The polarising pin engages with the polarising slot and ensures that the two halves of the connector are mated the right way round. The retention screws are tightened after the two halves of the connectors mate and this prevents them from separating due to shock or vibration.

Since a connector is part of an electrical circuit its resistance is an important parameter and is given in data sheets as its contact resistance. This resistance is usually specified at a certain current and is measured between the ends of the socket and pin so that it includes the resistance of the contact and that of the pin and socket materials.

Contact resistance is affected if a film of impurity forms over the contacts. They should therefore work with adequate pressure and a wiping action to remove this film. For clean contacts, the contact resistance is mainly due to the current crowding at the point of interface between the mating surfaces. If there is plastic deformation of the contacts then contact resistance is proportional to

$$R_C \propto \text{Material resistivity} \times \left(\frac{\text{Hardness of material}}{\text{Contact force}}\right)^{1/2}$$

If there is elastic deformation then for two contacts with equal radius the contact resistance is proportional to:

$$R_C \propto \text{Material resistivity} \times \left(\frac{\text{Modulus of elasticity}}{\text{Force} \times \text{Radius of contacts}}\right)^{1/2}$$

Therefore for low contact resistance the materials must have low resistivity, low hardness, low modulus of elasticity and must have a large radius and contact force.

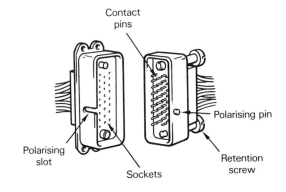

Fig. 6.23. A typical connector assembly.

The voltage rating of a connector is largely determined by the contact spacing and by the quality of the insulator material. The voltage breakdown paths between adjacent pins of a connector are through the air gap and the creepage path along the surface of the insulating material.

The insulation resistance of the connector is a measure of the leakage current flowing along the body of the connector and through the connector material. The leakage current will increase if there are contaminants on the surface of the insulator or in the material. The higher the leakage current the hotter the insulator becomes and this in turn reduces the insulation resistance and increases the leakage current.

The contact current rating of a connector depends on the dimensions of the contacts and also on the size of the wire connected to the contacts since these tend to carry away the heat generated in the contacts. To prevent hot spots occurring in the connector one should try to distribute the heavy current contacts as evenly as possible through the body of the connector. Hot spots reduce the maximum rating of the connector to a current less than that calculated from the rating of the individual contacts.

The mechanical mating and unmating forces needed to fit the two halves of the connector together are usually stated in data sheets. Connectors should have high contact forces once mated but the mating and unmating forces should be low. For large connectors having many pins these forces can be very high and other techniques, such as zero insertion force designs described in section 6.5.4, must be used.

Humidity, temperature and altitude are environmental factors which affect the performance of connectors. Moisture is absorbed by the insulator material and this increases its leakage current. A large amount of moisture on the surface of the insulator can also result in a short circuit between connector pins.

Extremes of temperature, both high and low, weaken metals and insulators, change the dimensions and the fit of the connectors, and crack seals and finishes. The life of a connector varies inversely with its *insert* temperature. An insert is usually a rigid material into which the contacts are mounted which allows them to move slightly and so helps

with alignment and reduces mating forces. The connector current must be derated at high ambient temperatures.

Both the inside and the outer case of a connector are cooled by air convection. At high altitudes this cooling effect is reduced since air density is lower, and therefore the connector current needs to be derated. When a connector is operated in a confined space, even at sea level, such that cooling by air convection is reduced, the effect is similar to that obtained by working at high altitudes, and the current must be derated.

At low atmospheric densities corona also occurs at lower voltages. This results in a reduction of the breakdown voltage rating of the connector with altitude.

**6.5.2** *Connector material*

The two main types of materials used in a connector are the metal for the contacts and the insulator for the body of the connector.

**6.5.2.1** *Contact material*

The contacts must act as electrical conductors and be able to withstand adverse mechanical and environmental conditions. They should have good spring qualities, be able to be formed accurately and cheaply, and they often need to have the capability of being plated. Four materials are usually used for contacts:

(*a*) Beryllium copper. This has good mechanical properties, high electrical and thermal conductivity, and good resistance to corrosion and wear. It has the best conductivity of the materials having comparable hardness, and it is possible to relieve all the internal stresses, caused during manufacture, by heat treatment. Beryllium copper is the most expensive of all the contact materials.

(*b*) Phosphor bronze. This has good corrosion resistance, fair conductivity, good flexibility, and is easily formed. It is used as a general purpose contact material up to a maximum operating temperature of 105 °C. It is twice as expensive as brass.

(*c*) Brass. This is a relatively cheap material and is used in applications where high temperature and high mechanical stresses do not occur. It can be reliably crimped, welded or soldered. The material loses conductivity with constant stress, and will not withstand many insertions and withdrawals. It also looses flexibility with age.

(d) Nickel–silver alloy. This material resists oxidisation and therefore does not always need to be plated. It has similar mechanical properties to phosphor bronze. Its properties degrade under constant stress although this degradation is less than that for brass.

Contacts are often plated as this usually extends their mechanical life, increases their resistance to wear, corrosion and chemicals, and lowers their contact resistance. A variety of plating materials is used, some of the most common being:

(i) Gold. Hard gold is usually plated onto the socket and soft gold onto the pin. This combination gives a burnishing action which improves the resistance to wear and reduces the contact resistance. Generally, gold plating is used when many insertions and withdrawals are needed, or when operating at low insertion forces, below about 100 gm, or at alow currents and voltages, below about 100 mA and 500 mV. It is under these operating conditions that oxides formed on the contact surface cannot be easily removed.

(ii) Gold overplating with nickel underplating. This is often used as it has excellent resistance to wear and hostile environments. The material has the surface characteristics of gold and since the nickel prevents migration of the base material into the gold, only a thin layer of gold is needed. Gold over copper is also used as it has low contact resistance, although it cannot stand many insertions and withdrawals.

(iii) Gold overplating with silver underplating. This gives low contact resistance and good resistance to wear and corrosion. The disadvantage of the system is that if the gold wears through or if it is very porous then the silver is exposed, and it is easily contaminated, especially in a sulphide atmosphere.

(iv) Silver. This is used on its own for high power contacts where high contact forces are used. It has poor shelf life and the silver tarnishes easily giving increased contact resistance, although this is not an important factor in power circuits.

(v) Rhodium and rhodium over nickel. This has very high resistance to wear and is used in high temperature applications. Its disadvantage is a high contact resistance.

Lubricants are often applied to contact materials to reduce their wear and increase their reliability. This has the added advantage that a layer of lubricant applied to a gold plated contact often allows the plating thickness to be reduced.

### 6.5.2.2 *Insulator materials*

The following properties are required from insulator materials used in connectors:

(i) Low water absorption, since this will reduce the insulation resistance and the breakdown voltage between contacts.

(ii) Resistance to abrasion. This minimises wear, which can cause misalignment of contacts and reduce pressure giving an increase in contact resistance.

(iii) Dimensional stability, to allow the sockets and pins to mate accurately.

(iv) Operation over a wide temperature range and no oxidisation, discolouration or embrittlement. These would set up stresses which could fracture the housing and give an unpleasant appearance.

(v) High tensile strength to prevent fracture when mechanically loaded.

(vi) The material must not ignite readily and must not emit poisonous gases. It should also be resistant to chemical cleaning fluids.

Two types of plastic materials are used in connectors as insulators, *thermoplastic* and *thermoset*. Thermoplastic can be remelted and resolidified. Thermoset cannot be remelted once set. Thermoplastic is more popular for use in connectors operating at temperatures up to about 300 °C since it is cheap, light and versatile. The initial cost of the material is higher than thermoset but the processing time is low and the scrap can be re-cycled. Thermoset plastic has a higher heat resistance than thermoplastic but it lacks the springy quality of thermoplastic. This means that it cannot be used in designs where the body material is used to form an internal spring catch to retain the plug and socket once they have mated.

The following thermoplastic materials are generally used for connector insulators:

(i) Polycarbonates. These have good electrical properties which are stable at high temperatures, humidity and frequency. They are mechanically strong and have good resistance to chemicals.

(ii) Silicones. These have excellent electrical properties which are stable with temperature, humidity and frequency. However the mechanical properties are not stable at high temperatures.

(iii) Polyphenylene oxides. These have high tensile strength and excellent electrical characteristics which are constant with frequency and humidity and over the temperature range $-40\,°C$ to $+150\,°C$.

(iv) Polysulphones. These have good electrical properties up to about $180\,°C$ and can withstand mechanical stress over long periods and at high temperatures.

Thermoset materials include the following:

(i) Epoxy. This is one of the most widely used materials due to the wide variety of types which are available, and the ease with which it can be used. The material exhibits good dimensional stability and low shrinkage.

(ii) Diallyl phthalates. These have high insulation resistance even at high temperatures and humidity. They are easy to mould and are capable of giving close tolerance housings. They are used for connectors operating in adverse environments.

(iii) Phenolics. There are many different types but they are all low cost and easy to mould. These are usually used where the environment is not very severe and the electrical requirements not very demanding.

Many different methods are used to make plastic housings.

(i) Compression moulding in which the moulding material is poured into hot open moulds having the same shape as the required part. The mould is closed and allowed to cool before the part is removed. This method is slow and requires preheating of the material. Flash is formed when the mould is closed forcing out some plastic, and this needs to be removed.

(ii) In transfer moulding the material is heated in a transfer pot and then forced under heat and pressure through an orifice into the closed mould. This method is faster than compression moulding since the cure time is shorter, but the cost of the mould is higher and waste material is left in the orifice and pot.

(iii) Injection moulding is the fastest method of moulding. The plastic material is fed into a cylinder where it is melted and then injected into the mould. This method gives good control over the melt temperature and produces a homogeneous mixture.

(iv) Extrusion is used to draw out from the melt long lengths of thermoplastic material having the same cross sectional area along its length.

(v) In vacuum forming, sheets of plastic are heated and vacuum is then used to draw the sheets against the sides of the mould. After cooling the sheets are removed and excess material is trimmed off.

(vi) In blow moulding the plastic is placed in a closed hot mould and air is injected into it at pressure. This forces the plastic against the sides of the mould and keeps it there until it cools.

### 6.5.3 Connector contacts

The contacts perform the most important function in a connector since they actually make the electrical connection. The contacts can be machined or stamped and they come in many shapes and sizes. Fig. 6.24 shows a pin and socket arrangement and this is very widely used as it is low cost and readily available. It has the disadvantage of being susceptible to damage and requires a high insertion force. The blade and fork arrangement shown in fig. 6.25 is generally used with printed circuit board connectors. It is low cost, reliable and has a low insertion force, but it is susceptible to damage, and dirt and humidity can accumulate in the connector cavity.

Many different socket arrangements are used for printed circuit board connectors where the edge of the board engages directly

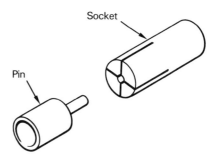

Fig. 6.24. Pin and socket contact.

with the socket. Fig. 6.26a shows a bellows arrangement where ribbon wire is formed to provide spring contacts which are deflected when the board enters the socket. This type of contact is low cost and has low insertion

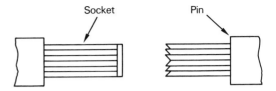

Fig. 6.27. Brush contact arrangement.

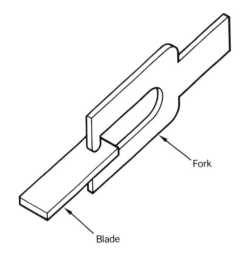

Fig. 6.25. Blade and fork contact.

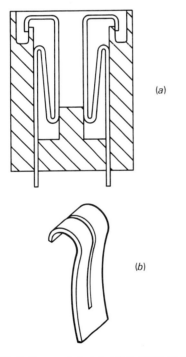

Fig. 6.26. Bellows socket; (a) socket arrangement, (b) bifurcated structure.

force, but it can be damaged and humidity can collect in the cavities. Fig. 6.26b shows a modification in which the wire is bifurcated to give a better contact area. The two halves of the spring are of different widths so that they have different resonant frequencies.

Fig. 6.27 shows an arrangement in which the pin and socket consist of two bundles of stiff wires which mesh with each other when in contact. This type of arrangement gives a good electrical contact at a low insertion force since several contacting prongs are used. Another contact arrangement which gives a very low insertion force is the hypertac contact shown in fig. 6.28. The socket has a number of wires which are formed at an angle on a tube wire carrier. A minimum of five wires are used. The pin deflects the wires to their elastic limit and they form separate connections along the length of the pin. This contact arrangement has very low contact resistance, high reliability, constant electrical and mechanical characteristics through the life of the contact, and very low insertion and withdrawal forces.

Many different techniques are used for maintaining the contacts in the body of the connector housing. Generally, the contact has protruding tines which are compressed during insertion but then fly out once they are in the contact cavity, and are retained by a shoulder in the moulding. Fig. 6.29 shows one such arrangement where the contact is inserted from B. A simple tool is used to remove the contact again by exerting force on the tines in the direction shown by the arrows at A so that they are compressed.

In some arrangements the contacts are a push fit into the housing and have a groove on the contact to help retain them, but there is no protruding tine arrangement. In other systems the contacts are moulded as part of the connector body and cannot be removed.

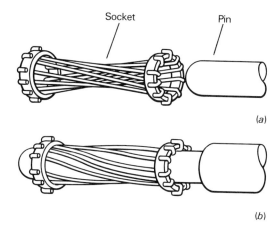

Fig. 6.28. The hypertac contact; (a) unmated, (b) mated.

Generally, four techniques are used to connect wire to the contacts.

(i) Crimping. This can be used with a variety of different wire types, stranded, coaxial or solid. The crimp gives a high pressure, gas tight connection and it can stand mechanical, chemical and environmental stress. No heat or flux or subsequent cleaning operation is needed. Automatic crimping tools are available so that skilled operators need not be used in manufacturing.

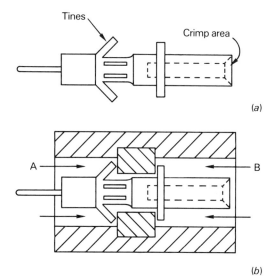

Fig. 6.29. Method of retaining contact in housing; (a) pin, (b) pin in housing.

(ii) Solderless wrap. This technique consists in wrapping wire tightly round the contacts which extend out from the back of the connector housing. It gives a good performance connection over long periods since the joint is gas tight. No flux or cleaning fluids are used and the joints can be unwrapped if required. Automatic wrapping machines are available so that the cost of making a joint is low. This system can only be used with solid wire and it needs a large contact area.

(iii) Solder or weld. Very good electrical connections can be obtained by these methods but skilled operators are needed. Quality control is difficult since visual techniques are not always reliable enough to indicate a bad joint. The system has high labour cost unless mass solder techniques are used such as solder dipping or wave soldering. A large contact area is needed for soldering but welding gives a high tensile strength joint from a small size, although it takes longer to make the joint and it needs more expensive equipment. The fluxes used in these systems must be subsequently removed or they can cause corrosion.

(iv) Insulation displacement connection. This is described in section 6.5.4.

### 6.5.4 Types of connectors

Connectors can be grouped by their applications, for example heavy duty, miniature, general purpose; or by their construction, for example rectangular, cylindrical, coaxial and D-type. In this section only a few special types of connectors, whose construction differs significantly from that shown in fig. 6.23, are described.

Fig. 6.30 shows the operating principle of the insulation displacement connector. It is mainly used with ribbon cable and has the advantage that the cable does not need to be stripped and then the wires individually connected to each contact. The sharp contacts pierce the insulation of the cable and make connections with all the conductors in one operation so that the time needed to connect wires to the contacts, called termination time, is greatly reduced. Fig. 6.31 shows one type of contact which can be used. The wire end of the contact must be sharp to cut through the wire insulation but it should not damage the strands of wire, therefore a material thickness

130  Peripheral components

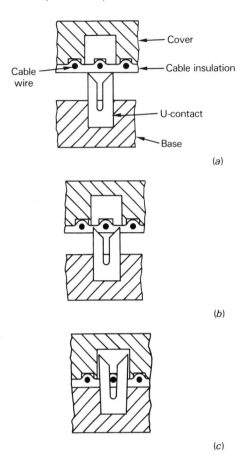

Fig. 6.30. Operation of an insulation displacement connector; (a) positioned, (b) piercing, (c) connected.

are usually high. In a two piece connector the pin half of the connector is soldered to the board so that the two mating halves are part of the connector assembly.

The material used in high voltage connectors must be able to withstand large voltages between terminals. Air can, in theory, withstand an electric strength of 20 kV/cm but in practice this is reduced to 5 kV/cm since it is difficult to maintain an equal field stress everywhere. Commonly used materials for high voltage connectors are epoxy resins, having a volume resistivity of $10^{14}$ $\Omega$ cm, PVC at $10^7$ $\Omega$ cm and PTFE with a resistivity of $10^{20}$ $\Omega$ cm. PVC is strong and cheap but it has a low melting point and is susceptible to attack by chemicals.

PTFE is one of the best materials for high voltage connectors since it has a high electrical resistance, good temperature stability and chemical resistance, and it is suitable for moulding in large volumes. The high voltage connector can be made as an aluminium body to which the screen of a coaxial cable is connected, and into which is formed a PTFE insert. This type of construction can be used for connectors usable at 15 kV to 20 kV. For voltages up to

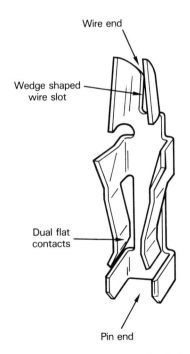

Fig. 6.31. Typical contact used in an insulation displacement connector.

of about 0.3 mm is used. The pin end must be springy and not overstressed when the pin is inserted so that thinner material, of about 0.15 mm thickness, is used. The slot at the wire end increases in width towards the inner end of the slot to reduce the stress on the contact after the wires have been pierced, and this reduces the risk of breaking the wires under cable flexing or vibration.

Printed circuit board connectors can be of one piece or two piece construction. In a one piece structure the edge of the board is plated and this acts as the pin half of the connector. Although this reduces cost, weight and space, careful control is needed on the production of the board with regard to board thickness and plating, and the insertion and withdrawal forces

## 6.5 Connectors

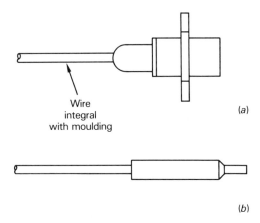

Fig. 6.32. A high voltage connector; (*a*) socket, (*b*) plug.

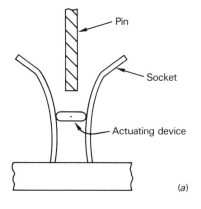

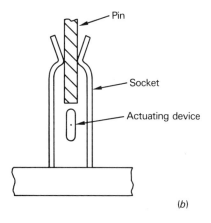

Fig. 6.33. Zero insulation force connector; (*a*) pressure relieved, (*b*) pressure applied.

about 50 kV the high voltage coaxial cable is moulded into the plug or socket in one piece as shown in fig. 6.32.

Connectors use spring pressure to keep the mating contacts in place. Zero insertion force (ZIF) connectors use a mechanism which relieves this pressure during periods of insertion and withdrawal. This is usually done by a lever or rotating device as shown in fig. 6.33. The actuating device separates the socket contacts during insertions and withdrawals and allows the socket to apply contact pressure once the pin is in place.

Integrated circuit sockets are a special form of connector which use the legs on the integrated circuit packages as the connector pins. These legs are primarily designed for soldering into boards and they can have faults such as plating corrosion and burrs which can cause connector (socket) problems. In common with other connectors low insertion and withdrawal forces are needed to prevent leg damage but high retention forces should be available to withstand board vibration. The sockets can bear onto the edge or side of the integrated circuit legs. In edge bearing sockets the contacts bear onto the sheer edges of the lead frame which have a rough finish. This gives poor electrical contact and the contact plating is degraded during insertion and withdrawal. Side bearing contacts press onto the flat side of the integrated circuit legs. Greater dimensional tolerances are permitted in the leads in this direction so the contacts are more difficult to design, but better electromechanical mating can occur since the contacts bear on the larger surface of the legs, and the surface is relatively smooth.

# 7. Quartz, ceramic, glass and selenium

## 7.1 Introduction

The components described in this chapter are primarily based on the less commonly used properties of quartz, ceramic, glass or selenium. For example glass envelopes used for reed relays are not dealt with here but switches based on the switching properties of glass are described.

Components made from quartz and ceramics use three effects exhibited by these materials, and these are the ferroelectric, piezoelectric and pyroelectric effects.

## 7.2 Ferroelectricity

Ferroelectricity is the electrical analogue of ferromagnetism. Ferroelectrical materials generally have no inherent electrical polarisation, but when they are subjected to strong electrical fields they acquire polarisation and are said to be 'poled'. The materials also exhibit a hysteresis loop in an alternating electrical field. Ferroelectric materials can be hard or soft. Soft materials are soluble in water, have a low melting point and are mechanically soft. Hard ferroelectric materials are not soluble in water, can withstand high temperatures and are mechanically hard. Hard materials are more frequently used for electronic components and are generally polycrystalline ceramics such as barium titanate, lead zirconate, cadmium niobate and lead lanthanum zirconate titanate (PLZT).

Ferroelectric materials are usually piezoelectric, as described in the next section. However, not all piezoelectric materials are ferroelectric, quartz being a notable example.

## 7.3 Piezoelectricity

### 7.3.1 *The piezoelectric effect*

The piezoelectric effect was discovered in 1880 by the Curie brothers. It relates mechanical stress in a crystal with an electrical signal. An electrical signal applied to a piezoelectric crystal causes it to be subjected to mechanical stress; or a mechanical stress applied to the crystal will generate an electrical signal across it. Piezoelectric materials are crystalline in structure and anisotropic in several properties. The same crystalline structure which gives the piezoelectric effect also gives the pyroelectric effect described in section 7.4. All pyroelectrics are piezoelectric but not all piezoelectrics are pyroelectric.

Piezoelectricity occurs in crystals which do not have a centre of symmetry. Twenty one classes of crystals do not have this symmetry and over one thousand crystal materials exhibit the piezoelectric effect. Single crystal and polycrystalline materials, such as ceramics, exhibit piezoelectricity. Piezoelectric ceramics are ferroelectric, and this enables them to be 'poled' so that the anisotropic properties necessary to piezoelectric action appear in a specific direction. Piezoelectric ceramics have no piezoelectric properties when first made due to the random orientation of their electric dipoles. Subjecting them to an electric field of between 10 and 30 kV/cm at a temperature just below the Curie temperature 'poles' the material so that it acts as a single crystal. The piezoelectric properties of bulk polycrystalline material are, however, not as good as those of true single crystal materials.

The advantages of piezoelectric ceramics over other piezoelectric materials are:
(i) they can be formed into complex shapes and still maintain uniform piezoelectric properties;
(ii) they are chemically inert and have a high Curie temperature;
(iii) they can be made using standard ceramic technology and the material composition and direction of polarisation can be varied relatively easily to meet any application requirements.

Single crystal quartz has a higher temperature stability and chemical inertness than piezoelectric ceramic and is ideal for oscillator frequency control applications. But to get good piezoelectric properties single quartz crystals must be accurately oriented and only perfect crystals must be used. The range of shapes available is also limited to simple structures such as plates and discs.

Depolarisation or depoling can occur in ceramic materials, with a loss of piezoelectric properties, under any of the following conditions:

(i) the material is exposed to a strong a.c. field, or to a d.c. field in a direction opposed to the original direction of poling;

(ii) the temperature of the material is not kept below the Curie temperature;

(iii) the mechanical stress on the material exceeds specified limits.

### 7.3.2 Piezoelectric equations

In an elastic non-piezoelectric material such as glass the relationships between an electric field $E$ and the resultant electric displacement $D$, and between a mechanical force $T$ and resultant strain $S$ in the material are given by

$$S = sT \quad (7.1)$$
$$D = \epsilon E \quad (7.2)$$

where $\epsilon$ is the dielectric constant or permittivity of the material and $s$ is its elastic compliance.

In a piezoelectric material the strain and electrical displacement depend on both the mechanical force and the electrical field and are given by the relationships

$$S = dE + s^E T \quad (7.3)$$
$$D = dT + \epsilon^T E \quad (7.4)$$

The superscripts $T$ and $E$ denote that $\epsilon$ and $s$ are measured at constant mechanical force and electric field respectively. $d$ is the charge per unit applied stress, at constant electrical field, or alternatively is the strain per unit applied field at constant stress. Therefore, $d$ is referred to as the piezoelectric charge constant of the material and is measured in units of metres per volt.

An alternative constant, the piezoelectric voltage constant $g$, is defined by the following two equations and is measured in units of volt metres per newton

$$E = -gT + D/\epsilon^T \quad (7.5)$$
$$S = s^D T + gD \quad (7.6)$$

The superscripts again denote the quantities which are kept constant. From (7.4) and (7.5)

$$d = g\epsilon^T \quad (7.7)$$

Therefore $d$ and $g$ can each be defined in two ways. $d$ can be defined for constant $E$ as the ratio of the resultant dielectric displacement to the applied mechanical stress, and measured in coulombs per newton. For constant $T$, the value of $d$ can be defined as the ratio of resultant strain to applied electric field, in units of metres per volt. $g$ can be defined at constant $D$ as the ratio of resultant electric field to applied mechanical stress measured in volt metres per newton, or at constant $T$ as the ratio of resultant strain to applied electrical displacement, and measured in units of square metres per coulomb.

The electromechanical coupling coefficient, $k$, of a piezoelectric device is defined by

$$k^2 = d^2/\epsilon^T s^E \quad (7.8)$$

or

$$k^2/(1 - k^2) = g^2 \epsilon^T/s^D \quad (7.9)$$

At low frequencies, below the frequency of mechanical resonance of the material, $k^2$ is a measure of how much energy supplied in one form, i.e. electrical or mechanical, is converted to the other form.

Typical values of $d$ and $g$ are $100 \times 10^{-12}$ to $600 \times 10^{-12}$ m/V and $10 \times 10^{-3}$ to $50 \times 10^{-3}$ V m/N respectively. The constants $d$, $g$ and $k$ are also dependent on the direction of the applied mechanical force or electrical field in the crystal. For example, fig. 7.1a shows the six possible directions where 4, 5 and 6 represent shear about the three axes. The constants are represented by two numbers such as $g_{31}$ where the first subscript is the direction of the generated field and the second is the direction of the applied mechanical stress. Fig. 7.1b shows an example where the electrical field and mechanical stress are about axis 3 and in fig. 7.1c the mechanical stress is about shear axis 4 while the electrical field is at axis 2.

134  Quartz, ceramic, glass and selenium

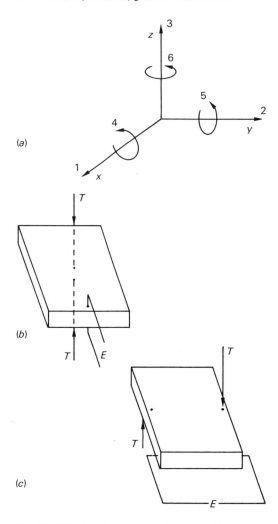

Fig. 7.1. Directional properties of piezoelectric constants; (a) axis rotations, (b) $g_{33}$, (c) $g_{24}$.

### 7.3.3 Piezoelectric and ferroelectric applications

#### 7.3.3.1 Oscillators

One of the main applications of piezoelectric quartz is as a control device in high stability oscillators. This application utilises the high $Q$ value of the quartz when in a mechanical resonant mode. The quartz frequency range is generally between 1 kHz to 750 kHz and 1.5 MHz to 200 MHz. It is difficult to cover the range between 750 kHz and 1.5 MHz due to strong secondary resonance. It is also preferable to use high frequency crystals, and count down to obtain lower frequencies, since the low frequency crystal is larger and less stable.

For operation in electronic circuits, the quartz crystal is carefully aligned using X-rays, and then it is precision cut along several planes and ground and lapped to exact dimensions. Electrodes are vacuum deposited onto the cut crystals and leads are attached to them and to the case of the crystals.

The quartz crystal has an atomic structure oriented about the $x$, $y$ and $z$ axes where $x$ is the electrical axis, $y$ is the mechanical axis and $z$ the optical axis. Many different cuts can be made to the crystal.

An $x$ cut is used for applications requiring mechanical change in thickness in the $x$ direction for an electrical field in the $y$ direction. It has a negative temperature coefficient of frequency. The $y$ cut is used for changes of length in the $y$ direction for an electrical field in the $x$ direction. It has a major face normal to the $y$ axis and exhibits a positive temperature coefficient of frequency. By taking a cut which is rotated from the $x$ and $y$ axes it is possible to combine the positive and negative temperature coefficients and get a cut which has negligible frequency shift for a small change in temperature. An AT cut is an example of such a cut which gives side face shear when an electric field is applied along the thickness direction.

Two cuts are most common for quartz controlled oscillators, the AT and BT cuts. The

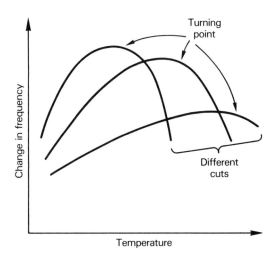

Fig. 7.2. Effect of crystal cut on its turning point.

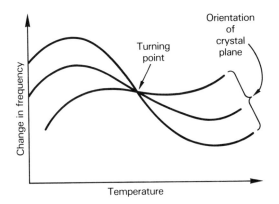

Fig. 7.3. Shift in frequency with temperature for an AT cut crystal.

AT cut has a lower variation of frequency with temperature and by operating it in a temperature controlled oven very stable oscillators are obtained. The BT cut is preferred above 10 MHz since it is less susceptible to drive levels and load capacitances.

All crystals have a frequency temperature characteristic similar to those shown in fig. 7.2. The position of the turning point varies with the cut of the crystal and can be chosen to occur in the range $-50\,°C$ to $+100\,°C$, to suit a given operating temperature range. The AT cut crystal has a cubic temperature curve, as shown in fig. 7.3, and shows a small change in frequency with temperature. By reorienting the crystal plane slightly it is possible to cause the curves to rotate about the turning point and so match any operating temperature range. Temperature controlled ovens are used to produce crystals for very stable frequency operation, and with a control of $\pm 1\,°C$ in the oven the stability of the crystal can exceed 1 ppm.

Fig. 7.4 shows the equivalent circuit of a quartz crystal assembly. $L_1$ represents the vibrating mass of the crystal blank and $R_1$ represents the loading on this crystal due to molecular friction and ambient air. $C_1$ is called the motional capacitance and represents the elasticity of the quartz material. $C_0$ is the static capacitance between the electrodes plated on the blank and stray lead and holder capacitance.

A quartz crystal can operate in a series resonant mode or a parallel antiresonant mode, and its impedance-frequency characteristic is similar to that shown in fig. 7.5. At the series resonant frequency $f_S$ the reactances of $L_1$ and $C_1$ are equal and the frequency is given by

$$f_S = 1/2\pi (L_1 C_1)^{1/2} \qquad (7.10)$$

The series resonant circuit then consists of $R_1$ in parallel with $C_0$. Above $f_S$ the inductive reactance increases and the capacitive reactance decreases. When $X_{L_1} - X_{C_1} = X_{C_0}$ parallel antiresonance occurs, at $f_P$, and the impedance is a maximum. The $Q$ value of the crystal is given by

$$Q = 2\pi f_S L_1/R_1 \qquad (7.11)$$

When the crystal is loaded any capacitance in the load will reduce $f_P$, and this effect can be

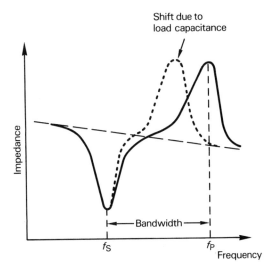

Fig. 7.5. Impedance characteristic of a quartz crystal oscillator.

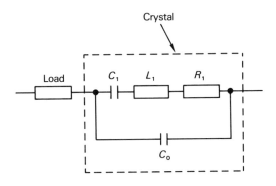

Fig. 7.4. Equivalent circuit of a quartz crystal.

used to give close tolerance in the operating frequency when it is in the antiresonant mode. The load capacitance is usually chosen as 30 pF above 1 MHz and 20 pF below 1 MHz. The ability to change the operating frequency of the crystal by varying the load capacitance is called the 'pullability' of the crystal and is used in applications such as voltage controlled and temperature controlled oscillators, and in crystal filters. The bandwidth of pullability is limited by $f_P - f_S$ since load capacitance reduces $f_P$. This is a function of $C_1/C_0$ and the $Q$ value of the crystal. $C_1/C_0$ varies over wide limits, between about $10^{-3}$ and $10^{-2}$ depending on the crystal cut, so that one can get narrow or wide bandpass characteristics.

Anything interfering with the mechanical vibration of the crystal will increase its equivalent series resistance $R_1$ and decrease $Q$. This can occur due to gas or air in the crystal enclosure, or due to acoustic reflections caused by supersonic waves generated by the vibrating crystal being carried by the gas to the enclosure walls and then being reflected back to the crystal. These effects can be avoided by evacuating the crystal enclosure. The crystal mounting structure is also part of the resonating system and must be carefully designed. Lead wire lengths must be in odd quarter wavelength increments to prevent out-of-phase resonance, which would interfere with the crystal's oscillations. Minute cracks and flaws on the crystal surface can also affect its oscillation, and too much solder on the lead attachment or poor plating adhesion or dirt will give crystal drag and increase its resistance.

In a quartz crystal, the frequency drift is asymptotic with time and therefore changes much more rapidly in the first few months of its life. This is called ageing and after this initial period the rate of change settles down to about 5 ppm per year in a temperature controlled environment. Ageing is due to contamination in the crystal, fatigue in the mounting wires and solder, outgassing of the materials inside the sealed holder, and leakage of the seals. If an oscillator or its oven is switched off for a few hours then the crystal recovers and the ageing process starts again at its initial rate, which is why a quartz standard should never be switched off.

### 7.3.3.2 *Filters*

Crystal filters are generally used where precise narrow band filtering is needed in the range 10 kHz to 100 MHz. They work on the same principle as oscillators and the passband frequency is determined by the cut of the crystal. Quartz is the material most commonly used for filters since it is stable and has relatively low cost, although it has a narrow bandwidth of about 0.5% of the centre frequency. Other materials such as lithium tantalate have a much wider bandwith. The minimum bandwidth is dependent on the $Q$ value of the resonator. AT cut crystals have a

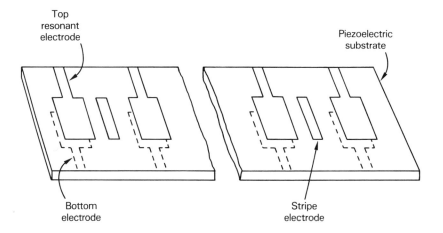

Fig. 7.6. A monolithic crystal filter.

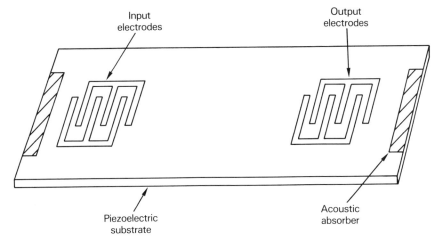

Fig. 7.7. A surface acoustic wave structure.

high $Q$ value but need careful design to avoid the loading problems described in the previous section.

The term crystal filter is applied to $LC$ filters having a crystal resonator. An alternative form of filter, called a monolithic crystal filter, has no external components. It consists of a piezoelectric substrate on both sides of which are deposited several pairs of metallic electrodes, as shown in fig. 7.6. The whole system acts as a mechanical filter in which the frequency of each resonant element is determined by the loading of the electrode and the finite plate boundaries. There is coupling between adjacent resonant electrodes and the stripe electrodes are used to adjust this coupling for high performance filters.

### 7.3.3.3 Surface acoustic wave devices

A surface acoustic wave (SAW) structure, shown in fig. 7.7, consists of a carefully polished and poled piezoelectric material on which are deposited input and output electrodes, usually of 0.01 μm thick aluminium in the form of interleaving fingers. The ends of the substrate are terminated in acoustic absorbent material to prevent coherent edge reflections. Applying an a.c. signal to the input electrodes sets up a series of mechanical surface waves in the piezoelectric material. These waves travel to the input electrode and are there converted back to an electrical signal. The backward travelling waves are absorbed in the acoustic absorber. The travelling wave is accessible over the whole length of the device so that it is easy to tap.

The characteristics of the SAW are primarily determined by the substrate material and the geometry of the electrodes. The width of the fingers and the gap between the fingers is usually a quarter wavelength of the acoustic wave which is to be preferentially excited. This means a dimension of about 8 μm at 100 MHz and 0.8 μm at 1 GHz. Because of these small dimensions a SAW has a planar structure and is usually made by photolithographic techniques similar to those used to make integrated circuits. The number of fingers used for the electrodes depends on the functional bandwidth required, and the overlap of the fingers is determined by the power of the signal source. The greater the overlap, the higher the proportion of input electrical energy which is converted to the acoustic mode.

Surface acoustic wave devices can be used in a variety of applications such as delay lines, bandpass filters, and oscillators. For an oscillator, the SAW is included in the feedback loop of an amplifier. This gives an oscillator with good short term stability, small size and a high fundamental frequency in the 10 MHz to 1500 MHz region.

### 7.3.3.4 Gas ignition

Piezoelectric materials are used to provide high voltages for lighters used in gas cookers, cigarette lighters, camping gas equipment, etc. They have a long operating life and are

138  Quartz, ceramic, glass and selenium

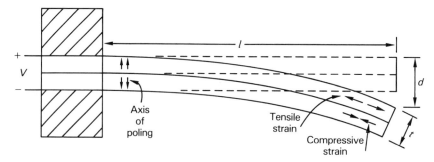

Fig. 7.8. A bimorph mounted as a cantilever.

generally used in the form of cylinders of the piezoelectric material. The cylinders must be made to exact dimensions and have parallel, square and flat end faces. A good coupling coefficient and permittivity are required, and the material must be mechanically strong and not easily depolarisable.

For an axially poled cylinder of length $l$ metres a stress of $T_3$ newtons per square metre in the direction of poling will give a voltage $V_3$ volts parallel to the direction of poling, where

$$V_3 \approx -g_{33} T_3 l \qquad (7.12)$$

$g_{33}$ is the piezoelectric voltage constant in volt metres per newton.

If $C$ is the capacitance of the unit at low frequency and $V_b$ is the breakdown voltage of the spark gap, then the maximum energy $E_M$ required for the gap is given by

$$E_M = \tfrac{1}{2} C V_b^2 \qquad (7.13)$$

The energy available per unit volume from the piezoelectric material is

$$\tfrac{1}{2} \epsilon_{33}^T g_{33}^2 T_3^2 \qquad (7.14)$$

Typical values for the materials used for gas ignition applications are $g_{33} = 25 \times 10^3$ Vm/N; $k_{33} = 0.7$; $\epsilon_{33}^T/\epsilon_0 = 1500$; peak open circuit voltage at $10^8$ Pa = 25 kV.

High mechanical stress in the material will cause depolarisation. Typical maximum values of static stress vary from $25 \times 10^6$ N/m² to $150 \times 10^6$ N/m². These values can be doubled by applying dynamic stress i.e. a short duration impact using a hammer and spring system.

### 7.3.3.5 Multimorphs and bimorphs

Multimorphs and bimorphs are flexure elements which operate in a bending mode in applications such as record player pick-ups, bell clappers, microphones, ultrasonic air transducers and small vibratory motors.

Fig. 7.8 shows the principle of a bimorph, which consists of two thin piezoelectric materials bonded together with their poling directions opposed. Voltage is applied to the outer ends and this causes one strip to expand lengthwise and the other to contract. This results in the cantilever being deflected by distance $d$.

A multimorph is made of one monolithic ceramic extrusion which works in the same way as a bimorph. Holes are formed in the centre of the ceramic and these are silvered to give a centre electrode and a top and bottom electrode, and these electrodes are used for poling the two halves of the device in opposite directions. In operation, only the top and bottom electrodes are used.

Fig. 7.9 shows the two methods of support commonly used with multimorphs. Typical parameters attained with these supports are

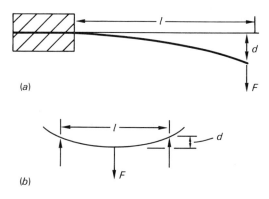

Fig. 7.9. Alternative mounting methods for multimorphs; (a) cantilever, (b) end-pinned.

| Parameter | Cantilever | End-pinned |
|---|---|---|
| Electrical charge output for force $F$ ($\mu$C/N) | $0.7 \times 10^{-3} \, l^2$ | $0.2 \times 10^{-3} \, l^2$ |
| Electrical charge output for deflection $d$ ($\mu$C/mm) | $6/l$ | $20/l$ |
| Deflection $d$ for applied voltage; $F = 0$ (mm/V) | $7 \times 10^{-7} \, l^2$ | $2 \times 10^{-7} \, l^2$ |
| Force $F$ for applied voltage $d = 0$ (N/V) | $5 \times 10^{-3}/l$ | $2 \times 10^{-2}/l$ |

Fig. 7.10. Typical characteristics of a multimorph.

given in fig. 7.10. The voltage output can be found from the charge divided by the total capacitance of the multimorph plus any circuit shunt capacitance. The maximum capacitance of the multimorph is about $60\,l$ pF where $l$ is in millimeters. The maximum bending force to prevent depoling is about $1.5 \times 10^{-3}$ N m and the maximum applied voltage before which depoling occurs is about 200 V.

#### 7.3.3.6 Piezoelectric display

Fig. 7.11 shows the principle of one type of piezoelectric display. Light inside the glass which strikes the glass–air interface at angles greater than the critical angle will be reflected back into the glass. From the viewing side this interface will therefore seem to be opaque. If now another material having a refractive index similar to glass is put within a fraction of a wavelength of light away from the glass then, even for angles greater than the critical angle, the light is not reflected but goes through the new media. From the viewing side the glass–air interface is now transparent.

A piezoelectric display utilises this effect by placing a piezoelectric material, with the required character format, close to the glass–air interface. An electric signal applied selectively to the different parts of the material will cause it to move towards or away from the glass–air interface. The display can have the appearance of clear characters on a dark background or vice versa.

A piezoelectric display does not emit its own light but reflects the ambient, so in this respect it is very similar to a liquid crystal display. This means that the display does not suffer from washout effects in strong ambient light. The piezoelectric display is much faster than a liquid crystal display, having a response time of about $10\,\mu$s and it can operate over a wide temperature range. The display does not need any polarisers so that it has no viewing angle problems. The disadvantages of piezoelectric displays are that they need a high operating voltage, in the region of 100 to 200 V d.c. and that they have many moving elements, giving a high assembly cost.

#### 7.3.3.7 Electro-optic applications

The display described in the previous section utilises the piezoelectric effect. Ferroelectrics exhibit a truly electro-optical effect which results in a change of the order of 1% in refractive index induced by an applied electric field.

PLZT is the most promising electro-optical material. It can be made to have a wide range of electrical and optical properties by varying its composition. It has a large electro-optic effect, and is optically transparent when cut into thin plates and polished. The scattering mode of operation of PLZT has advantages over other methods needing polarised materials since the light losses associated with the polariser are minimised, the system is simpler, and the plate thickness and polarisation are no longer critical parameters. The disadvantage of this

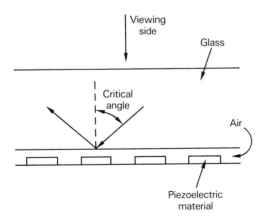

Fig. 7.11. Principle of piezoelectric displays.

mode of operation is that there is less contrast.

Other ferroelectric materials used for electro-optic applications include gadolinium molybdate and bismuth titanate. Gadolinium molybdate is also ferroelastic, which means that it exhibits a stress-strain hysteresis loop. It is used as a page composer for laser holography. Bismuth titanate crystals can have their domain patterns detected optically, and have high resolution capability. Patterns can be written into and erased from these crystals and they can store patterns and be read many times before erasure. They are used for optical memories and displays.

Several devices are being developed which use the electro-optic properties of ferroelectric materials. These include optical modulators, light deflectors, displays and optical memories. The ferroelectric display depends for its operation on the change in polarisation in thin plates of PLZT ceramic, brought about by an applied electric field. The material is optically transparent and contains small grains with about ten domains per grain. The domains are randomly polarised but a poling field will cause them to be aligned. A second control field between two transparent electrodes can cause the vectors to be rotated by 90°. The light is rotated within the ceramic, and if it is placed between two polarisers then, depending on the orientation within the two polarisers, the light can be made to pass through the unit either when the ceramic is energised or deenergised. The transparent electrodes on the ceramic are shaped to the display formats required.

This display arrangement is capable of high quality images and it is non-volatile so that it can also be adopted for memory applications. The display can be completely erased by the poling field or selectively erased by a light beam in conjunction with a photoconductive film. The disadvantage of the display is that it needs a separate poling field and poling electrodes, and the PLZT material is relatively expensive.

## 7.4 Pyroelectricity

Pyroelectricity is the development of opposite charges on the end faces of a crystal which is subjected to temperature change. It results from the stress set up in the crystal due to temperature gradients. If a crystal develops a positive charge on one face during heating it will have a negative charge there during cooling. The charge is gradually lost if the temperature is kept constant.

The amount of polarisation or charge developed on the crystal is proportional to the change in temperature. The crystal must be eccentric and must belong to a class of crystal having particular types of symmetries. The pyroelectric effect is generally small but in some materials, such as triglycine sulphate, it is large enough to be used in thermal sensing systems.

## 7.5 Glass

### 7.5.1 *Principles of switching glass*

Amorphous material such as glass has the potential for replacing traditional semiconductor materials like silicon in certain applications. Amorphous material is non-crystalline in structure, but exhibits many of the properties of crystalline semiconductors like silicon. It is metastable since it prefers to be in the crystalline rather than the non-crystalline state. In switching glass the material is made to switch between the crystalline and non-crystalline states.

Switching glass is available as a chalcogenide, which is a compound of sulphur, selenium and tellurium, and as an oxide, which is closer to the structure of common glass, to which various compounds such as the oxides of copper or vanadium may be added.

The rate of phase change in glass is dependent on the stability of the glass and this in turn is determined by its composition. The composition can be changed by adding elements with different bonding characteristics. The stability of the glassy state is also determined by its temperature. For stable glass the rate of change to a crystalline phase is very low over a wide temperature range, but for low stability glass the phase change is much more rapid at high temperatures. This rate of phase change determines the switching characteristic of the devices based on glass.

Bulk glass needs a high voltage for switching and it switches relatively slowly and has a short operating life. Thin films are used in practical devices and these may be formed by vacuum

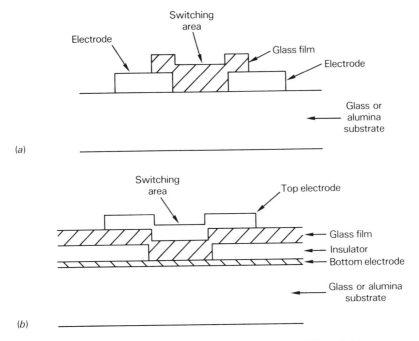

Fig. 7.12. Glass switching structures; (*a*) planar, (*b*) sandwich.

deposition or by screen printing. The films can be made in many ways although two fundamental structures are mainly used. The first is the planar arrangement shown in fig. 7.12*a*. Electrodes are deposited close together on a substrate of glass or alumina and the glass film is deposited between them. Switching occurs at the surface of the glass film and is therefore affected by surface conditions. This structure does not therefore give devices with reproducible characteristics.

The second type of device structure is the sandwich arrangement shown in fig. 7.12*b*. The bottom electrode is deposited on the substrate and onto this a glass film of about 1 μm thickness is vacuum deposited followed by the top metal electrode. The insulating film has a circular hole in the middle which is used to define more clearly the switching area. This prevents switching from occuring at the edges of the electrodes, which can result in failure, and it also gives devices with reproducible characteristics.

Gold is normally used as the electrode material since it resists oxidisation and has high electrical and thermal conductivity, and will not damage the glass layer when deposited on it. Sometimes a refractory film, such as molybdenum, is put between the electrodes and the glass film to prevent the electrode material from diffusing into the glass.

### 7.5.2 *Application of switching glass*

Glass can have the characteristics of a memory or a threshold switch. The threshold switch has *monostable* properties and switches from a high impedance non-crystalline state to a low impedance crystalline state at a given voltage. Therefore in fig. 7.13*a* the device is normally in a high impedance state having a resistance of about $10^5$ Ω, and very little current flows as the voltage across it is increased. When the voltage reaches the threshold value, which is in the region of 25 to 30 V, the device switches to a low impedance state, with a resistance of about 100 Ω. The device will now remain at this low impedance as long as the current is maintained above a holding value of about 0.2 to 0.5 mA. The glass switch generally operates better under pulsed rather than d.c. conditions.

When a voltage greater than the threshold value is applied to the glass switch there is a time delay before it switches to a low

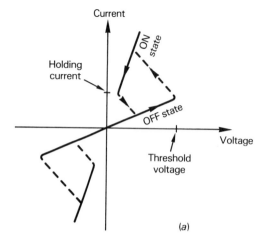

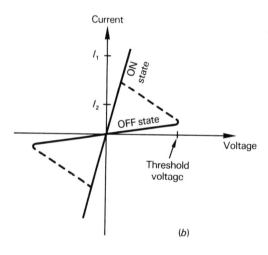

Fig. 7.13. Switching glass characteristics; (a) threshold switch, (b) memory.

impedance. This delay is of the order of several microseconds but it decreases as the overdrive voltage increases, reaching a few nanoseconds at twice the threshold voltage. The switching time, after the delay period, is of the order of picoseconds. The threshold switch has a life of about $10^8$ to $10^9$ operations.

The actual switching mechanism in glass is not fully understood. The switching current usually flows in a narrow filament through the material between the electrodes. This causes high power dissipation and heat generation which results in an alteration in the atomic spacings within the material and accounts in part for the switching action.

The glass device operating as a memory switch has the properties shown in fig. 7.13b. In the high impedance non-crystalline state it has a resistivity of about $10^7$ $\Omega$ cm and it can be switched into the crystalline state, of resistivity about 100 $\Omega$ cm by a voltage pulse of about 25 V. The threshold voltage is generally in the region of 10 V to 20 V depending on the thickness of the glass film, the temperature, and the pulse parameters.

The switching operation is similar to that in a threshold switch, and the memory switch will behave in a monostable mode if the threshold voltage is removed within a certain time. To keep the memory switch in the low impedance mode a current of about 10 mA is passed through the device for about 10 ms immediately after it is switched. This current results in the transition of the material from the non-crystalline to the crystalline state.

To return from the low impedance to the high impedance state requires a relatively high current pulse, of about 200 mA for 5 ms. This causes some or all of the crystalline bonds to be disrupted so returning the material to the non-crystalline or high impedance state. The shape of the pulse drive used to switch from a high to low impedance state is critical since it must first have a sufficiently high voltage to cause the device to go to a low impedance and then a sufficiently large current, usually between $I_1$ and $I_2$, must flow for a long enough time to ensure that the device is stable in the low impedance state. The closer the value of the set current to $I_2$ the less the reset current needed to convert the device back to the high impedance state. The memory switch can be read by passing a low current through it, of the order of 2.5 mA, and noting the volt drop. The state of the memory cell is not changed after a read operation.

The glass switch can also be used as a temperature sensor. The threshold voltage is stable to within 1% at constant voltage, but it decreases linearly with temperature at the rate of 1% to 2% per degree centigrade in the range 0 °C to 70 °C. This property can be used in temperature sensing circuits.

## 7.6 Selenium

Selenium was discovered by Berzelius in 1817. It is extracted from sulphurous minerals and exists in five forms only two of which are of importance for electronic components. In the amorphous state it looks black and glassy and at high temperatures it is converted into a metallic grey substance with a melting point of about 220 °C.

The main applications of selenium for electronic components is in rectifiers and as *transient voltage* suppressors. Selenium rectifiers are robust and have negligible reverse recovery time since their charge storage is small. They are gradually being replaced by silicon but are still used in applications such as EHT supplies for television where they compete with silicon rectifiers on cost and on their capability to withstand transients without damage.

Selenium for rectifiers is made by adding metal and halide additives to molten selenium at about 400 °C. The melt is then rapidly chilled by casting on a cold surface and is then stripped off, ground and sieved to give selenium powder of the correct particle size.

The selenium is now vacuum evaporated or hot-pressed onto a nickel-plated base plate, as shown in fig. 7.14. Holes are punched into the base plate for mounting and it is then put into an oven, which changes the amorphous selenium into a semiconducting hexagonal crystalline state. A counter-electrode is then sprayed or deposited onto the surface of the selenium. This electrode is made from alloys of tin and cadmium, and can include additives such as bismuth and thallium. It has a very low melting point of about 150°C to 180°C and so it can be sprayed onto the selenium surface relatively easily. A barrier layer, consisting of a lacquer coating usually based on cellulose nitrate or nylon, is sometimes added between the selenium layer and the counter-electrode. This separates the additives in the selenium from the cadmium in the counter-electrode.

The metal base plate in the selenium rectifier plate provides mechanical support only, and is usually made from aluminium alloy or steel. Aluminium is preferred since it has better thermal conductivity and is more corrosion resistant. The rectifying action occurs between the selenium layer and the counter-electrode.

The selenium plates initially have a very low reverse voltage blocking capability. They are electroformed by passing a reverse current through them for a few hours and gradually raising the applied voltage until it reaches the required value. The plates are then cut to size and are assembled on insulated steel spindles, or small plates are packed in plastic, ceramic or paper tubes. Each plate can handle about 30 V rms and a forward current of 300 mA/cm$^2$. Its threshold voltage is about 0.6 V, the maximum operating temperature is 80 °C, and the operating life is about $10^5$ hours. Selenium plates can be stacked in parallel or series to give higher voltages and currents, or for various arrangements such as bridge, push-pull, half-wave and full-wave. The stacks are painted or lacquered by immersion or spraying and this protects them from humidity and corrosive atmospheres. Precautions are needed to keep mercury vapour away from selenium plates as this can drastically reduce their blocking capability.

If a selenium rectifier is operated for a long time in a humid atmosphere, or operated continuously in the forward direction only, it

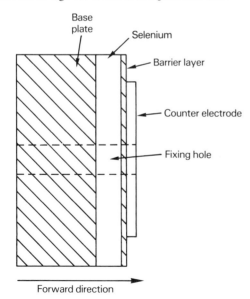

Fig. 7.14. Cross section of a selenium rectifying plate.

can lose its voltage blocking capability. The plates can be reformed to their original value. Small puncture spots can also appear in the plates due to melting of the counter electrode at these points. This causes the crystalline selenium to melt and solidify after cooling into the amorphous state, so that the punctured spots isolate themselves, and since they are small they do not significantly affect the characteristics of the selenium rectifiers.

Selenium plates can be built to have well defined and sharp reverse breakdown characteristics, and these are used as voltage transient suppressors. Each plate has a voltage limit of 50 to 100 V and several plates can be connected in series to increase this voltage. Plates can also be connected back-to-back to give bidirectional protection. The transient energy is dissipated in the selenium plates and the amount of energy which can be dissipated is determined by the plate size. Commercial devices range from 5 mm to 100 mm square and they can carry currents in excess of 1000 A for about 10 ms.

# 8. Power sources

## 8.1 Introduction

Power sources can be divided into *primary cells* and *secondary cells*. A primary cell is used once in its life time, until it is spent, and is then discarded. It cannot be recharged and used again. Examples of this type are carbon-zinc, alkaline-manganese, mercuric oxide, silver oxide, zinc chloride, zinc-air and lithium.

Secondary cells need to be charged after assembly before they can be used. However if the cell is spent it can be recharged and used again, and one can go through many such charge-discharge cycles during the life of the cell. Examples of secondary cells are lead-acid, nickel-cadmium and silver-zinc.

Cells produce an output voltage which is a function of the electrochemistry of the material used within the cell. Several cells can be combined within one package to produce a larger output voltage and this package is called a battery.

The primary and secondary cells mentioned earlier are the ones most commonly used in electronic equipment. This chapter also describes the more specialised power sources such as fuel, solar, reserve and standard cells.

## 8.2 Cell characteristics

Perhaps the characteristic of most interest in cell design is the voltage of the cell. The *open circuit voltage* depends on the state of charge of the cell and on its recent history. Fig. 8.1. shows how this voltage can vary with time depending on whether the cell has recently been charged or discharged.

The cell voltage also varies with the amount of energy which it has provided, as shown in fig. 8.2. Generally the discharge curve is relatively flat until the cell is almost completely discharged and then it decays rapidly. The fall in cell voltage is also more marked at low temperatures.

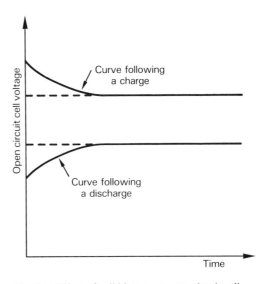

Fig. 8.1. Effect of cell history on open circuit cell voltage.

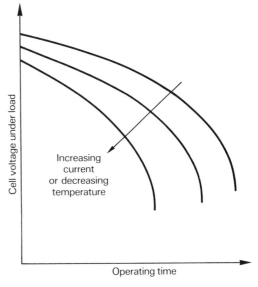

Fig. 8.2. Effect of load or temperature on the cell voltage, at constant temperature.

The amount of energy which the cell can deliver without its voltage falling below a specified value is referred to as the cell's capacity, and is usually measured in units of ampere hours. This is also often stated in multiples of its '$C$' rate. A fully charged cell which is discharged at its '$C$' rate will be fully discharged in one hour. So a cell which has a capacity of 2 ampere hours has a '$C$' rate of 2 amperes. If it is operated at a $0.2C$ rate then the cell will be discharged after 5 hours, while at the $5C$ rate it will be discharged in 12 minutes.

An alternative measure of the cell's capacity is given by the integral of the ampere hour product over the time needed to discharge the cell from the open circuit voltage to some cut-off voltage. This is measured in units of watt hours or joules. Both the ampere hour and watt hour ratings of the cell are dependent on the operating temperature, the period of storage prior to use, the discharge rate, and the discharge duty cycle.

The energy density of a cell is the ratio of cell energy output to its weight or volume, and is measured in watt hours per kilogram or watt hours per cubic centimetre. The discharge rate of the cell is the current at which the cell is discharged, usually expressed as a ratio of its rated capacity. The *depth of discharge* is the percentage of the rated capacity by which the cell has been discharged. So a 100 ampere hour cell which has been discharged by 20 ampere hours has a depth of discharge of 20%.

The charge rate of a secondary cell is the charging current expressed as a function of the cell's rated capacity. For a secondary cell the cycle life is the total number of charge–discharge cycles which the cell can withstand before failure occurs. This failure can be total, such as an internal short circuit, or a partial failure such as the reduction of the cell's capacity below a specific value.

The charge acceptance is the ability of a secondary cell to accept energy. It is measured as the proportion of the charge input which the cell can give out again without discharging below a specific value. This is usually less than 100% and depends on several factors such as the state of charge of the cell, the charge rate, the amount of overcharge, the environment and the cell history. The charge acceptance increases as the charge rate increases or if the cell temperature is lowered or the state of charge is reduced.

The charge voltage is the voltage developed in the cell while it is under charge. During constant current charging this voltage can be higher than the rated discharge voltage by up to 50%. The charge voltage increases with the charge rate and is reduced at high cell temperatures.

The *charge retention* of the cell is its ability to hold charge when it is stored. This is also referred to as the self-discharging property of the cell, and it is dependent on the temperature. It varies from 0.01% per day for a nearly discharged cell at 0 °C to 2% for a fully charged cell at 40 °C. The *shelf life* of the cell is usually stated as the period of time at a specified temperature after which the cell has reached a defined percentage of its original capacity.

The internal resistance of the cell is generally responsible for causing the output voltage droop on load, and in giving power losses in the battery. The resistance is inversely related to cell size and increases at low temperatures and so restricts the charge–discharge capacity of the cell. It also depends on other factors such as the state of charge of the cell and its age.

The mechanical characteristics of a cell are just as important as its electrical parameters. The case is designed to conduct away the maximum amount of heat generated in the cell, and the cell must be mounted to help in this heat conduction. The maximum allowable size of the cell is also limited in many applications, and mobile applications may require the cell to be able to withstand shock, vibration and accelerating forces without any fluctuations in the output voltage, and to be operated in any position.

## 8.3 Primary cells

### 8.3.1 *A basic cell*

The structure of a simple cell is shown in fig. 8.3. Two dissimilar metal plates are immersed in an electrolyte consisting of an acid, an alkali or a metal salt. The metal plate which is more electropositive reacts chemically with the electrolyte and gives off positive ions, so that the plate is negatively charged. The positive ions move to the other plate and give it their charge so that this plate acquires a positive charge. The charges on the plates re-

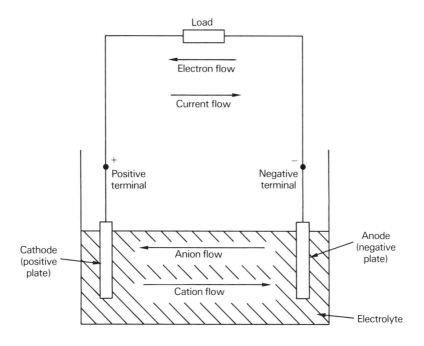

Fig. 8.3. Operation of a basic cell.

sult in a potential difference between them and a current will flow if a load is connected between the two plates. The value of the cell potential difference is determined by the plate material and not by its size. The amount of current the cell can deliver depends on the area of the plates, and the energy content of the cell depends on the quantity of the electrons released and accepted, and this is determined by the volume or weight of the materials used.

When a current flows in an external circuit the positive ions in the electrolyte are neutralised at the cathode by combining with an electron, and this forms a neutral 'film' around the positive plate. This film lowers the charge on the plate and hence the potential difference of the cell. It also increases the internal resistance of the cell and reduces the current flow. The cell is said to be polarised in this state. In practical cells some form of depolarisation is needed which reacts with the neutralised ions and so removes the neutral film. Depolarisation is not generally required in secondary cells which can be recharged to regenerate them. Some cells use self-depolarisation in which the neutralised ions are chosen so that they are of the same material as the positive plate, so that the film builds up the positive plates and does not affect the flow of ions.

**8.3.2** *The carbon-zinc or Leclanché cell*

This cell was one of the earliest to be developed and is still the most widely used, mainly due to its good overall performance, availability and low cost. It is shown in fig. 8.4 and consists of an outer casing of zinc, which acts as the negative electrode, and a central carbon rod which is the positive electrode. The carbon rod is surrounded by a mixture of manganese dioxide and powdered carbon contained in a porous sack. This acts as a depolariser and neutralises the hydrogen film, which is formed around the carbon rod during cell operation, turning it into water. The electrolyte is a mixture of ammonium chloride and zinc chloride dissolved in water. The cell is called a dry cell since the electrolyte is made in paste form.

The neutral hydrogen film around the positive plate forms faster than it can be removed by the depolariser, so that if the carbon-zinc cell is operated in a continuous mode it rapidly looses its e.m.f. The depolariser con-

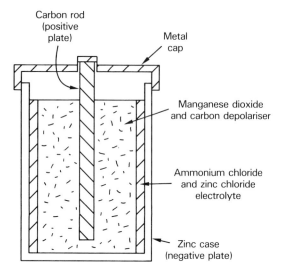

Fig. 8.4. The carbon–zinc or Leclanché cell.

tinues to work even after the cell is off so the cell is best suited for applications needing an intermittent duty cycle. Fig. 8.5 shows the voltage–time graph for a cell operating in an intermittent duty cycle. The terminal voltage recovers to a lower voltage during rest periods as the cell's ampere hour capacity is gradually reduced.

The carbon–zinc cell has an optimum operating temperature of 20 °C to 30 °C. Its terminal voltage and capacity are severely reduced below 10 °C. It also suffers from local chemical action between the electrolyte and the electropositive metal, even when the cell is in storage and not drawing any current. The degree of local action depends on the impurities present in the metal and is worse at high temperatures. This gives the carbon–zinc cell a relatively low shelf life. The cell also has a tendency to leak if left in a discharged condition for long periods. This is caused by the electrolyte leaking out and corroding the zinc container.

### 8.3.3 *The zinc chloride cell*

The zinc chloride cell is very similar in structure to the Leclanché cell with the exception that the electrolyte contains zinc chloride only whereas the electrolyte in the Leclanché cell is a mixture of zinc chloride and ammonium chloride. The ammonium chloride in the Leclanché cell prevents oxygen from the air reacting with the zinc chloride and so gives the cell a longer shelf life. However, ammonium chloride also inhibits electrochemical action and reduces the cell's capacity. Leaving out ammonium chloride gives the cell a much higher current capability and minimises polarisation. To prevent action with air the zinc chloride cell has better seals than the Leclanché cell.

The zinc chloride cell uses water during

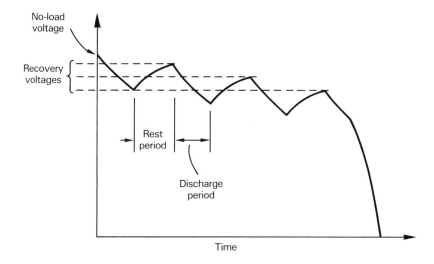

Fig. 8.5. Voltage recovery in a carbon–zinc cell operating on an intermittent duty cycle.

operation so that it is almost dry at the end of its life. The battery has good leak resistance and can operate at temperatures well below freezing.

### 8.3.4 *The alkaline manganese dioxide cell*

The alkaline manganese dioxide cell uses an alkaline electrolyte of potassium hydroxide or sodium hydroxide instead of the acidic electrolyte used in the Leclanché cell. This allows the cell to have a higher energy to volume ratio, a better high current performance and a longer shelf life, although it is more expensive.

The alkaline manganese dioxide cell uses an 'inside-out' structure compared to the Leclanché cell since the positive plate (cathode) is connected to the outer case. The cathode is a mixture of manganese dioxide and graphite which is compressed to fit around an anode made of zinc pellets mixed into a paste with the electrolyte. The anode and cathode mixtures are separated by layers of electrolyte absorbent material. The cell is contained in a steel case which is in contact with the cathode.

The no-load terminal voltage of the alkaline manganese dioxide cell is 1.5 V and its cost and capacity is between those of the Leclanché and mercuric cells.

### 8.3.5 *The mercuric oxide cell*

The structure of the oxide cell is shown in fig. 8.6. The cathode is made from zinc, which can be either a foil or pressed into shape. The anode is mercury which is liberated from the mercuric oxide during operation, and the electrolyte is a concentrated aqueous solution of potassium hydroxide and zinc oxide. These react to form potassium zincate which prevents the caustic solution from attacking the zinc electrode and emitting hydrogen, and this gives the mercuric oxide cell its high stability. A small percentage of graphite is generally added to the mercuric oxide to reduce the internal resistance of the cell.

In operation zinc ions are emitted from the negative electrode and enter the electrolyte to give off hydrogen ions. These ions move to the mercuric oxide and combine with oxygen to form water and displace positive mercury ions which are neutralised by electrons. The cell is self-depolarising since mercury, which accumu-

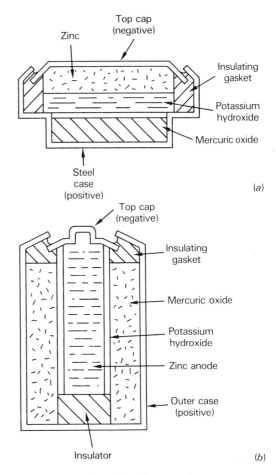

Fig. 8.6. Mercuric oxide cell construction; (*a*) button cell, (*b*) cylindrical cell.

lates on the positive electrode, is the electrode material.

Two types of mercuric oxide cells are available commercially. One has an open circuit terminal voltage of 1.35 V and a very flat voltage–current discharge curve. It is used mainly as a reference source. The second type of cell has manganese dioxide added to the mercuric oxide which gives it an open circuit terminal voltage of 1.4 V. The voltage–current discharge curve is less flat than that of the 1.35 V cell and the battery is mainly used in consumer applications which can tolerate this.

The mercuric oxide cell has an energy to volume ratio which is three times that of the Leclanché cell, and a low internal resistance which is maintained until the end of life. The constant voltage of the mercuric oxide cell is primarily due to its self-depolarisation capability. The cell can be short circuited or heavily overloaded and it will return to its full open circuit voltage within a few minutes.

A mercuric oxide cell has a much longer life than the Leclanché cell, with a smaller change in capacity with temperature. The long shelf life of the mercuric oxide cell is primarily due to the reduction in its local action. The cell has an optimum storage temperature of 21 °C but it can be stored over the range -20 °C to +100 °C.

Mercuric oxide cells are generally used in applications needing a stable voltage supply, long life, high capacity and small size. They are more expensive than Leclanché cells.

### 8.3.6 *The silver oxide cell*

The construction of the silver oxide cell is very similar to the button mercuric oxide cell. The cathode is depolarising silver oxide, the anode is zinc and the electrolyte is highly alkaline. The voltage of the cell is 1.5 V and this higher voltage, compared to the mercuric oxide cell, is needed in some applications.

Monovalent silver oxide ($Ag_2O$) has generally been used for the cathode and this gives a cell with an energy density about the same as the mercuric oxide cell but more expensive. Newer types of cells using divalent silver oxide (AgO) give 40% to 50% more energy than earlier types. Divalent silver oxide is metastable and oxygen liberated during cell operation combines with it so that it is suitable for use in sealed cells.

### 8.3.7 *The lithium cell*

Fig. 8.7 shows the construction of a lithium cell. It consists of a spirally wound core in a steel case. Lithium is used as the anode and it has the highest electrode potential, the highest ampere hour per unit weight, and the highest melting point of all alkali metals. Many different cathode systems can be used and this determines the open circuit cell voltage as follows: thionyl chloride (3.6 V), vanadium pentoxide (3.4 V), silver chromate (3.0 V), sulphur dioxide (2.9 V), copper sulphide (2.2 V).

The lithium–sulphur dioxide system is relatively cheap and uses a lithium anode and a gaseous sulphur dioxide cathode dissolved in acetonitrile on a carbon conductor. It is a high

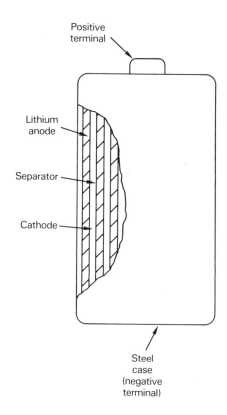

Fig. 8.7. Lithium cell.

pressure system and so the cell must be designed with safety vents which can release corrosive sulphur dioxide. The lithium–thionyl chloride system is the most popular. It uses a lithium anode and a gaseous cathode dissolved in organic electrolyte. This gives a low pressure system so the cell can be hermetically sealed, and is therefore better for high temperature applications.

The electrolyte used is non-aqueous so that its conductivity at low temperatures is much better than other cells, and at -55 °C it can still provide 50% of the cell's rated capacity.

The cell can operate over very wide temperature ranges and has a shelf life of over 5 years at 20 °C, which is much longer than other types of cells. The voltage–current discharge curve is almost flat and the capacity of the cell is ten times greater than that of the Leclanché cell and four times more than that of the mercuric oxide cell. Lithium is expensive and it will react violently with even small traces of water. Very tight seals must therefore be used although carefully designed safety vents are still required to allow gas to escape in the event of a short circuit.

Lithium cells are generally used in applications needing very long life or operation at extremes of temperature.

### 8.3.8 *The zinc-air cell*

The zinc-air cell uses zinc as the anode material and oxygen from the air as the cathode material. This means that the space and weight normally taken up by the cathode-oxidising agent can now be allocated to the anode material giving this cell a high energy to weight ratio. The key to the use of the zinc-air cell is the material called polytetrafluorethylene (PTFE or Teflon) which lets air into the cell but does not allow the electrolyte to leak out.

Two cell structures are shown in fig. 8.8. The flat cell has been used for several years in railway and marine signalling applications, but the miniature button cell is a more recent innovation. In the flat cell the anode is amalgamated zinc powder and contains the electrolyte which is a concentrated solution of potassium hydroxide. The cathode is made in layers held in a plastic frame. The outermost layer is a microporous PTFE film which allows the

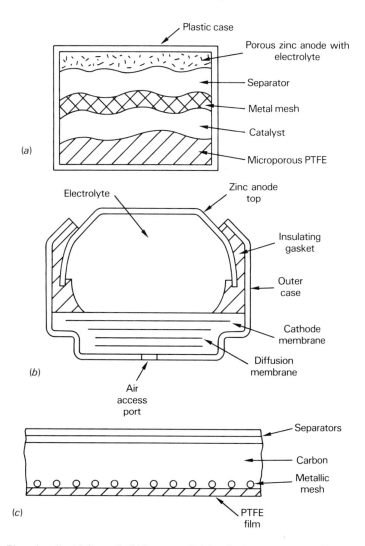

Fig. 8.8. Zinc-air cells; (*a*) flat cell, (*b*) button cell, (*c*) cathode arrangement of button cell.

oxygen from the air to come into contact with the electrolyte but prevents the electrolyte from escaping. The second layer is a catalyst which is in contact with the electrolyte and helps in the chemical reaction and results in a high current density at the cathode. The metal mesh layer acts as the positive terminal connection and collects current generated in the cell. A permeable separator prevents direct electrical contact between cathode and anode but allows free passage of ions in the cell. The cell can be used mounted in any position.

In the button cell the anode is the zinc top and the potassium hydroxide electrolyte is contained within it. An insulating gasket separates the anode and cathode halves of the button. The air cathode arrangement shown in fig. 8.8c is only 0.5 mm thick. It is made up of a system of separators on the side facing the electrolyte. The carbon provides the catalytic action needed to combine the oxygen from the air with the hydroxide from the electrolyte to give water. The metallic mesh provides mechanical support and acts as the current conductor and the PTFE film allows air to enter the cell but prevents electrolyte from escaping.

During operation of the zinc-air cell the zinc reacts with the electrolyte and leaves electrons on the anode giving it a negative charge. The chemical reaction is

$$Zn + 2OH \rightarrow ZnO + H_2O + 2e^-$$

At the same time the air-cathode structure helps with the reaction of oxygen at the surface of the electrolyte. Oxygen from the air combines with water from the electrolyte, drawing electrons from the cathode structure and replacing the hydroxyl ions lost at the anode. Electrons removed from the cathode leave it positively charged. The chemical reaction is

$$\tfrac{1}{2}(O_2) + H_2O + 2e^- \rightarrow 2OH^-$$

The current capability of the cell depends on the rate of air flow and on the cathode surface area. The ampere hour rating of the cell depends only on the weight of the zinc anode. The cell reaches the end of its life when all the zinc material has been used up and even after this point the air cathode can still be used many times over.

The zinc-air cell has a no-load voltage of 1.4 V, a high energy to weight or volume ratio and a higher current than the alkaline or mercuric oxide cells. It is used in applications requiring continuous or long periods of operation at high currents. The cell has low internal resistance, which is mainly determined by the oxygen diffusion rate. It can operate over the wide temperature range of -40 °C to +60 °C. The zinc-air cell has excellent storage capabilities since one of the cell's reactants is oxygen and this can easily be excluded by an air tight wrapping around the cell. The average loss in capacity of the cell during storage is only about 2% per year.

The zinc-air cell is inherently safer than the lithium or mercuric oxide cells since it has a built in safety vent via the air diffusion path, and its short-circuit current is limited by the oxygen absorption rate of the cell. The main problem with the zinc-air cell is that it can be affected by atmospheric conditions. The water-vapour pressure in the cell is equivalent to 55% relative humidity at 20 °C so that on dry days the cell will lose moisture, and gain it on wet days. This affects the aqueous electrolyte, which is usually 30% potassium hydroxide, so that the cell can fail if operated at extremes of atmospheric conditions over long periods. Atmospheric carbon dioxide can also react with the electrolyte to give potassium carbonate which increases the internal resistance of the cell.

## 8.4 Secondary cells

### 8.4.1 *The lead-acid cell*

The lead-acid cell was developed by Gaston Plante in 1860, but it was its application much later in automobiles that resulted in its widespread use. The disadvantage of the automobile cell type of construction is that there is risk of spillage, the cell needs frequent topping up of the electrolyte and it must be mounted upright. For electronic applications the gelled-electrolyte cell is used which is sealed, can be mounted in any position and does not need topping up with electrolyte.

In the gelled-electrolyte cell both plates of the cell are made from lead which is in the form of a grid. The positive plate is filled with lead dioxide and the negative plate with spongy lead. These plates are made in the form of thin metal sheets and are interleaved with layers of highly porous fibreglass separators, and wound into a compact cylinder. This cylinder is then sealed in a chemically stable polypropylene case and put into a metal case for added strength.

The gel type electrolyte is close to the metal plates and since these are thin, and a spirally wound construction is used, most of the active material of the metal plates is kept to the surface. This gives the cell a low internal resistance, low polarisation and long life. The separator material is chosen to withstand heat and oxidation. It separates the positive and negative plates electrically, and acts as a wick to hold the electrolyte and to distribute it evenly over the working area.

The electrolyte is a water solution of sulphuric acid. The quantity of electrolyte is chosen so that it can be retained by the plates and the separators, and none of it is free to leak. This also gives good gas diffusion and oxygen recombination so avoiding large gas pressures in the cell during overcharge. However a safety vent is usually provided to allow for the escape of gas.

During the discharge period the lead dioxide in the positive plate and the spongy lead in the negative plate react with the sulphuric acid to form lead sulphate and water. Charging reverses the process. During overcharge no more lead sulphate is available and now the charging current electrolyses the water in the electrolyte and forms oxygen and hydrogen gas at the positive and negative plates, respectively. This can lead to a rapid build-up of gas pressure in the cell. To prevent this the cell is usually designed such that the hydrogen formed at the negative plate is suppressed during charging. The oxygen formed at the positive plate moves to the negative plate and is absorbed there by the lead. The hydrogen then combines with the oxygen to form water, so that any large amount of gas evolution is suppressed except during periods of severe overcharge.

The lead–acid cell has a nominal cell voltage of 2.1 V, which is the highest of the secondary cells, and has the lowest cost per watt-hour. It can withstand high charge and discharge rates and is specially good on pulsed operation where it can deliver high capacity in a short time. During the rest periods acid diffuses from the separator back to the working areas of the plates and allows a greater working capacity.

The cell voltage droops during the discharge period due to loss in the internal resistance of the cell. There is also a drop in capacity at higher discharge currents because of insufficient ion diffusion caused by the depletion of electrolyte near the active material of the plates. This effect is called concentration polarisation.

The lead–acid cell will operate over a temperature range of -60 °C to +60 °C although its optimum operating temperature is about 20 °C. The cell capacity and the terminal voltage decrease with temperature due to a reduction of the ionic diffusion rate, with a voltage–time relation similar to that shown in fig. 8.2.

The shelf life of the lead–acid cell is relatively long compared to other types of secondary cells. The shelf life is reduced by internal electrochemical discharge, which is worse at high temperatures. The loss of charge per day varies from 0.01% for a nearly discharged cell at 0 °C to 2% for a fully charged cell at 45 °C.

The chemical effects of self-discharge during storage are similar to those which occur during normal operation, and the end result is the formation of lead sulphate. However the crystals formed during self-discharge are large and completely surround the active material particles. This process is called sulphation and if the crystal growth is allowed to continue long enough it can prevent the cell from accepting charge. In a completely self-discharged cell the lead sulphate dissolves in the very dilute sulphuric acid electrolyte and then diffuses into the separator. When the cell is subsequently charged the lead sulphate changes to metal lead and forms tracks in the separator, which eventually short circuits the cell. To avoid sulphation problems the lead–acid cell should be stored at low temperature and recharged during storage; it should never be stored in a discharged state.

### 8.4.2 *Nickel–cadmium cell*

The nickel–cadmium cell was first developed by Waldemar Jungner in 1900. In the discharged state the positive electrode of the cell contains nickel hydroxide and the negative electrode contains cadmium hydroxide. Charging converts nickel hydroxide to a more oxidised state and cadmium hydroxide to cadmium. The action of the cell is as follows. At the negative electrode absorption of electrons occurs and cadmium hydroxide is reduced to cadmium:

$$Cd(OH)_2 + 2e^- \rightarrow Cd + 2OH^-$$

This is followed by a side reaction:

$$2H_2O + 2e^- \rightarrow 2OH^- + H_2$$

The hydrogen charging potential is not reached when the cell is being charged at moderate rates so the side reaction which evolves hydrogen does not occur to any large extent.

The following action occurs at the positive electrode:

$$2Ni(OH)_2 + 2OH^- \rightarrow 2NiO(OH) + 2H_2O + 2e^-$$

followed by a side reaction:

$$4OH^- \rightarrow 2H_2O + O_2 + 4e^-$$

The oxygen formed on the positive electrode is encouraged to migrate to the negative plate by its close proximity. Here the oxygen is absorbed by the same process which absorbs electrons:

$$O_2 + 2H_2O + 2Cd \rightarrow 2Cd(OH)_2$$

No hydrogen is evolved as long as the rate of oxygen evolution at the positive plate is not greater than its rate of usage at the negative plate. Most cells contain a surplus amount of cadmium hydroxide at the negative plate to allow for some overcharging. This is called a charge reserve and results in the negative plate not being fully charged when the positive plate is fully charged.

The electrolyte used in the nickel–cadmium cell is potassium hydroxide, but it does not play any direct role in the cell's action. Water is produced on charging and is absorbed on discharge, and the purpose of the electrolyte is to reduce the effect of water on the cell's operation.

Two types of constructions are used for nickel–cadmium cells, cylindrical and button. Cylindrical cells are more versatile and can tolerate more abuse during usage than button cells. The electrodes are made by depositing nickel or cadmium hydroxide into flexible porous plates which are formed by sintering fine nickel powder onto a nickel substrate. The sintered plates are then separated by a nylon or polypropylene separator, which must be porous, stable and inert, and capable of retaining the electrolyte. The assembly is then wound and sealed in a nickel plated steel case.

In a button cell, plates of nickel and cadmium are used, separated by a porous separator, and then stacked. The greater the number of electrodes in the stack the lower the internal resistance of the cell and the larger its discharge current capability. The electrodes in the button cell are usually made by compressing the active material into metal mesh pockets, in plates. This gives the cell a relatively large storage capability. Gas pressure safety vents are not provided in the case so the button cells must be operated under carefully controlled conditions and should not be charged by currents greater than about $0.01C$. There is generally a progressively greater loss of cell capacity with each charge–discharge cycle.

Cylindrical nickel–cadmium cells can usually accept continuous charge, in the region of $0.1C$, without deterioration, and they are often used in standby applications where they are on continuous charge. The cells have very low internal resistance so that they should not generally be connected in parallel to give greater current since one cannot reliably predict the extent of current sharing. Most cylindrical cells are designed to vent at pressures of 10 to 20 atmospheres.

The end point voltage of the nickel–cadmium cell is about 1.0 V, at which point it is almost exhausted. Its capacity is generally stated as the discharge rate which will bring the cell voltage to 1.0 V in 5 hours, that is $C/5$. The capacity of the cell varies with the discharge rate, as shown in fig. 8.9 and with temperature, as shown in fig. 8.10. Although the cell will operate over a temperature range -40 °C to +60 °C the optimum operating temperature is 20 °C to 30 °C. At low temperatures the internal resistance of the cell increases so that the voltage droops at high discharge rates. Temperature effects are more pronounced on charging than on discharging. The ability of the cell to recombine with oxygen during overcharging is reduced at low temperatures, so cells should not be charged at very low temperatures. The oxygen recombination reaction is also exothermic and causes the cell temperature to rise. Charging a cell at temperatures greater than 40 °C will usually severely

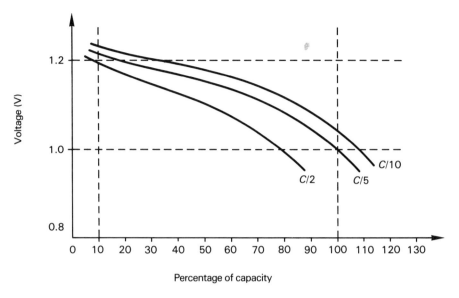

Fig. 8.9. Variation of capacity of a nickel–cadmium cell with discharge rate.

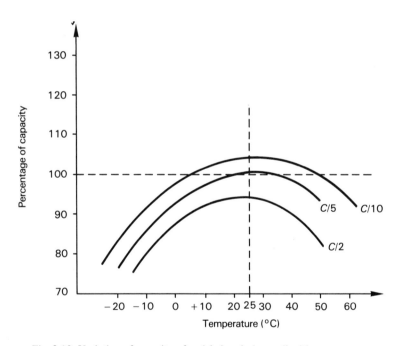

Fig. 8.10. Variation of capacity of a nickel–cadmium cell with temperature.

limit its life and in these instances it is better to charge at room temperature and then warm the cell up to its operating temperature.

The nickel–cadmium cell does not show any marked voltage increase at the end of its charging cycle so a constant current charging system, rather than constant voltage, should be used. The efficiency of charge is about 80% making it necessary for 20% more energy to be put into the cell than is taken out. The cell can be stored indefinitely in a charged or discharged state without any loss of life. The cell will become discharged due to self-leakage but it can readily be recharged. However, the cell exhibits a 'memory' effect such that if it is stored in a discharged state or operated at a very low discharge rate over a long period then it 'forgets' its original capacity and becomes conditioned to operate at this lower capacity level. It can be reconditioned by putting it through several charge–discharge cycles of gradually increasing capacity.

### 8.4.3 *Silver–zinc cell*

The silver–zinc cell is a comparatively new development. It is made in a sealed button form as shown in fig. 8.11. The positive electrode is silver and is in the form of a wire mesh which is coated with silver oxide by an electrochemical process. The negative electrode uses a substrate consisting of a screen or perforated metal plate made of silver or silver plated copper. This is then coated with zinc oxide either by electroforming or by applying it as a paste and then drying and firing. Each electrode in the silver–zinc cell is saturated in the electrolyte of potassium hydroxide.

The separator shown in fig. 8.11 is usually made of cellulose material. Several problems have been encountered with this material and new materials are under development. The cellulose separator is hydrated by the electrolyte and can expand and press against the zinc anode so that it is not as readily available for dissolution. The cellulose separator can also be oxidised and will degenerate giving a short circuit between the positive and negative electrodes.

The silver–zinc cell has more than double the capacity of a nickel–cadmium cell of the same size, and it has an almost flat discharge curve. Its terminal voltage is 1.5 V so that it needs fewer cells in a battery stack than a nickel–cadmium cell. During charging the terminal voltage of the silver–zinc cell rises rapidly to 1.85 V when the cell reaches 95% of its charged capacity, and this enables it to be used with simple charging systems.

The silver–zinc cell does not exhibit a memory effect like the nickel–cadmium cell. It has a 95% charge efficiency and this is very useful when it is being charged from sources such as solar cells. It also has a low leakage current and is often used as a back-up system to drive low current electronic circuits.

The disadvantages of the silver–zinc cell are its limited cycle life (of about 100 cycles), and its high cost, primarily due to the cost of silver.

### 8.4.4 *Battery charging*

Many different charging methods and charging circuits are used for secondary batteries, but the principle of all of them is shown in fig. 8.12. The d.c. source voltage need not be very smooth and usually the unfiltered output from the a.c. mains will suffice. The resistor $R$ includes the internal resistance of the battery, and of the source. The charging current is then given by

$$I = (E_s - E_b)/R \qquad (8.1)$$

where $E_b$ is the battery voltage and $E_s$ is the supply voltage.

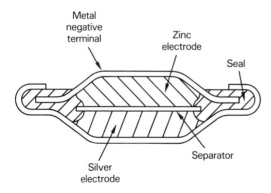

Fig. 8.11. The silver–zinc cell.

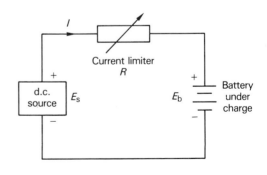

Fig. 8.12. Connection for charging a battery.

In the constant voltage charging methods the source voltage is a little above the voltage of the fully charged battery and $R$ is kept constant. When the battery is discharged the value of $E_b$ is low and the charging current is high, but this current gradually tapers off as the battery voltage increases. This charging method is also known as taper charging and fig. 8.13a shows the charging current waveform. In order to reduce the power requirements of the charger the initial current may be limited, as shown in fig. 8.13b.

Constant voltage charging can be used very effectively for batteries such as lead-acid and silver-zinc, which show an increase in voltage towards the end of their charge period. Nickel-cadmium cells have an almost flat charge characteristic, so a constant voltage charging system results in a heavy current flow, even after the battery reaches its fully charged state, and this causes oxidisation and overheating. Constant current charging methods, which are simple and cheap, are preferred for these cells. Referring to (8.1) if the supply voltage $E_s$ and current limiting resistor $R$ are large then the slight variation in battery voltage between the discharged and charged states will not appreciably affect the charging current. Some chargers use a stepped system which reduces the voltage as the battery nears its charged state to avoid overcharging.

Constant current charging is often used to trickle charge batteries on standby duty. As these batteries will be on charge a long time before they are used the charging current can be kept low to reduce heating.

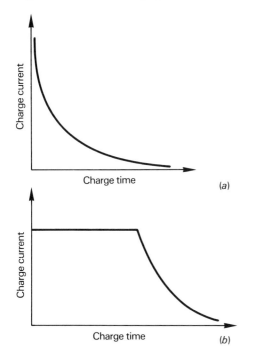

Fig. 8.13. Charging current from a constant voltage charging system; (a) no current limit, (b) with current limit.

## 8.5 Battery selection

If the application requires a battery with low initial cost then primary batteries, which are

| Parameter | Carbon-zinc | Zinc chloride | Alkaline manganese dioxide | Mercuric oxide | Silver oxide | Lithium | Zinc-air |
|---|---|---|---|---|---|---|---|
| Nominal cell voltage (V) | 1.5 | 1.5 | 1.5 | 1.35 or 1.40 | 1.5 | 3.0 | 1.4 |
| Energy output (watt hour/kg) | 40 | 90 | 60 | 100 | 130 | 300 | 200 |
| Shelf life at 20 °C (years) | 1.5 | 2 | 3 | 3 | 3 | 5 | 5 |
| Operating temperature range (°C) | +5 to +60 | −10 to +60 | −10 to +60 | −20 to +100 | −10 to +80 | −40 to +80 | −40 to +60 |
| Flatness of discharge curve (comparative) | 3 | 3 | 3 | 2 | 2 | 1 | 1 |
| Cost (comparative) | 4 | 3 | 3 | 2 | 2 | 1 | 1 |

Fig. 8.14. Comparison of primary cells. In the comparative data 1 = highest or best, 4 = lowest or worst.

thrown away after use, are the obvious choice. Secondary batteries can prove cheaper in the long run, over the life of the equipment, since they can be recharged and re-used. However, the need to provide a battery charger with the equipment puts up the initial cost. Secondary batteries are also not preferred if the life of the equipment is short.

Fig. 8.14 compares some of the parameters of primary cells. For general purpose use the choice is usually from carbon–zinc, zinc chloride, mercuric oxide, alkaline manganese dioxide or silver oxide. Lithium and zinc–air batteries are used for special applications only.

The carbon–zinc cell is cheap and readily available and is a good choice for applications having a light, intermittent duty cycle. It does not have a long shelf life or a very flat discharge curve. For heavy duty, continuous, high current applications alkaline manganese dioxide batteries are preferred. They have a longer shelf life and about 50% to 100% more energy for the same weight than carbon–zinc cells. They are not as economic as carbon–zinc for current drains below about 200 mA.

The mercuric oxide battery has a high energy to weight ratio and a flat discharge curve. It is often used in voltage reference applications. The silver oxide battery has characteristics similar to mercuric oxide but with a higher nominal cell voltage. Lithium batteries are used in applications needing very high energy densities or a wide operating temperature range or a long shelf life.

Fig. 8.15 compares the parameters of secondary cells. Lead–acid cells cost between one half and one third as much as nickel–cadmium cells, and this difference is more noticeable in larger sized batteries. Lead-acid cells also have a higher voltage so that fewer cells are needed in a battery. This results in a lower chance of failure, or defects due to cell imbalance and low cell capacity.

Nickel–cadmium batteries are used in applications where many charge–discharge cycles are involved, and where the discharge curve is to be flat. They also have a lower volume and weight than lead–acid batteries, and below about 0.5 ampere hour capacity the nickel–cadmium cell becomes cost competitive with lead–acid cells.

### 8.6 Fuel cells

The fuel cell works on the principle that the chemical energy obtained from the oxidation of a fuel is converted directly into electrical energy. The earliest cell of this type to be developed was the hydrogen–oxygen system shown in fig. 8.16, where the oxygen may be obtained from the air. The cell reactions are as follows:

At the anode:
$$H_2 \rightarrow 2H^+ + 2e^-$$

At the cathode:
$$O_2 + 4H^+ + 4e^- \rightarrow 2H_2O$$

Overall:
$$2H_2 + O_2 \rightarrow 2H_2O$$

This results in a difference of potential between the two electrodes and a flow of current in a load connected between them.

A fuel cell works isothermally and it can have a very high thermal efficiency, in the region of 60% to 80%. In a low temperature cell the electrodes are made of finely divided platinum in the form of wire screens and the electrolyte

| Parameter | Lead-acid | Nickel-cadmium | Silver-zinc |
|---|---|---|---|
| Nominal cell voltage (V) | 2.1 | 1.2 | 1.5 |
| Energy output (watt hours/kg) | 20 | 30 | 110 |
| Shelf life at 20 °C (years) | 1.0 | 0.2 | 0.3 |
| Cycle life (number of operations) | 500 | 2000 | 100 |
| Operating temperature range (°C) | −60 to +60 | −40 to +60 | −20 to +80 |
| Flatness of discharge curve (comparative) | 2 | 1 | 1 |
| Cost (comparative) | 3 | 2 | 1 |

Fig. 8.15. Comparison of secondary cells. In the comparative data 1 = highest or best, 4 = lowest or worst.

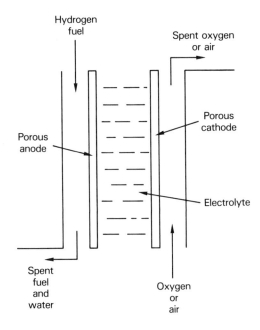

Fig. 8.16. Operation of a hydrogen–oxygen fuel cell.

is potassium hydroxide (KOH). Fig. 8.17 shows the characteristics of such a cell. The power density measured in milliwatts per square centimetre is greater when pure oxygen is used since it is a better oxidant than air.

There have been many developments of fuel cells. For example some cells have used hydro-

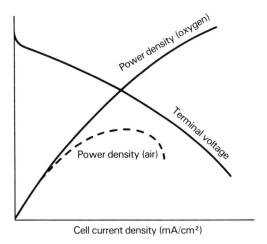

Fig. 8.17. Characteristics of a hydrogen–oxygen fuel cell.

carbon gases or methanol which is cheaper than hydrogen. These cells use a catalytic reformer which produces carbon monoxide and hydrogen, and then changes carbon monoxide to carbon dioxide in a second catalyst before passing the hydrogen into the fuel cell. Methanol is a good fuel since it has a low carbon content, it is cheap, and it can be easily handled and stored. The chemical reactions of the cell are:

At the anode:
$$CH_3OH + H_2O \rightarrow CO_2 + 6H^+ + 6e^-$$

At the cathode:
$$O_2 + 4H^+ + 4e^- \rightarrow H_2O$$

Using a liquid electrolyte presents several problems in fuel cells. Porous walls are needed to hold the solution, and the chemical reaction occurs in the capillaries of the walls where there is contact between the gas, solution and solid. It is difficult and expensive to make these walls. The key to future developments in fuel cells is continuing research into the replacement of the solution by a solid. With solid electrolytes, the surface used is no longer critical and cheaper catalysts, higher temperatures, and less close control of temperature can be used. The solid electrolyte must conduct oxygen ions or protons very quickly.

With a solid electrolyte the cell can be run at higher temperatures, and this gives larger cell voltages. The power to weight ratio is also improved by a factor of three since a very thin layer of electrolyte can be used. In some present cells, zirconia doped with lime is used as the electrolyte. This has high ion conductivity at 1000 °C but this falls off at lower temperatures. Noble metals are used as the electrodes due to the oxidising conditions involved.

Fuel cells have primarily been used in aerospace and military applications. However, research is currently going on into forms of cells using solid electrolytes at room temperature. These have great potential for use in thin film batteries, which can be formed onto a printed wiring board for electronic equipment.

## 8.7 Solar cells

### 8.7.1 *Principle of solar cells*

The Earth receives energy from the Sun, which on a bright day corresponds to a power density of about 1 kW/m² at the latitude of the United Kingdom. If 10% of this energy could be harnessed it could provide the entire present power demand of the world.

Two systems exist for harnessing the Sun's energy directly. In the first the energy is converted to heat, and in the second it is converted directly into electricity. The second effect is called photovoltaic and it is this system which is described here.

The photovoltaic cell was introduced in section 2.4.2. It consists of a pn junction as shown in fig. 8.18a. Light falling on this cell will generate hole–electron pairs which are swept away to the two electrodes and produce a voltage between them. The amount of energy needed to release electrons is called the band gap of the material.

The electrical characteristics of a photovoltaic cell are shown in fig. 8.19. When the cell is not illuminated it behaves like an ordinary diode. If illuminated it generates a voltage $V_0$ when no current is taken from the cell, and generates a short-circuit current $I_s$ which is the reverse current without any applied voltage. $I_s$ increases in proportion to the intensity of light falling on the cell whereas $V_0$ increases less slowly, approximately as the logarithm of the light intensity. The maximum value of the open circuit voltage is close to the band gap energy of the material.

The theoretical maximum power output from the cell is the product $V_0 I_s$ but due to the curve of the characteristic the practical limit to the power output is given by $V_M I_M$. The fill factor or curve factor of the photovoltaic cell is defined by the ratio:

$$\text{Fill factor} = V_M I_M / V_0 I_s \qquad (8.2)$$

If $I_p$ is the intensity of the incident light, measured in milliwatts per square centimetre then the efficiency of the cell is given by

$$\text{Efficiency} = V_M I_M / I_p \qquad (8.3)$$

Photovoltaic cells have efficiencies in the region of 5% to 25%. There are several losses which account for this low figure. First, only that part of the incident light which has the correct wavelength and sufficient energy can release hole–electron pairs, and for silicon cells and solar light this is 77% of the light at sea level. The lower the band gap of the cell material the greater the amount of the incident light used.

Some of the incident light is reflected, as shown in fig. 8.18, and some is absorbed by the top electrodes. The amount of light reflected depends on the relative refractive index between air and the cell material. The lower the refractive index of the cell the less the reflection. Generally solar cells are covered by a thin coating of antireflective material which reduces reflection losses, although this coating absorbs some of the incident light and prevents it reaching the junction of the cell.

Losses also occur due to some of the incident

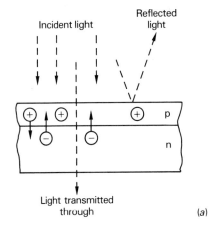

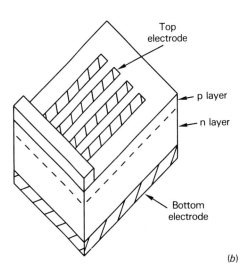

Fig. 8.18. A photovoltaic cell; (a) schematic, (b) construction.

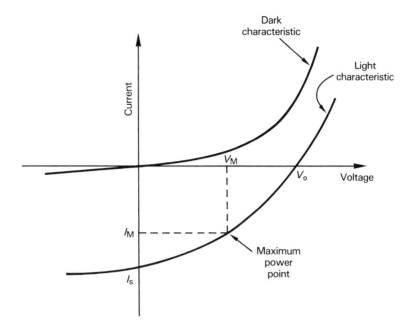

Fig. 8.19. Photovoltaic cell operating curve.

light going right through the cell and not generating hole–electron pairs. The amount of light transmitted depends on the thickness of the cell and the absorption coefficient of the material at the wavelength of the incident light.

A further drop in efficiency occurs due to resistance losses since current must flow along the thin top layer of the cell before reaching a top electrode. Increasing the area of these electrodes will reduce the resistance loss but will also cut down on the amount of light falling on the cell.

Holes and electrons which are liberated in the photovoltaic cell due to thermal energy combine with the photo-emitted hole–electron pairs and give rise to a further loss within the cell. This is called the junction loss and it increases with temperature. The higher the band gap of the cell material the less the portion of the incident light energy which goes to heat the cell since most of the energy is required to release hole–electron pairs. The increase in loss with temperature is also lower for semiconductors with higher band gaps.

Photogenerated hole–electron pairs can recombine on their own before reaching the electrodes. Generally this recombination occurs at defects in the crystal lattice so that very pure material is required for high efficiency solar cells.

### 8.7.2 Silicon solar cells

Silicon has a band gap of 1.12 eV and gives a cell with maximum open circuit voltage of 0.6 V, but under full sunlight and in high ambient temperature this falls to about 0.4 V.

The efficiency of a silicon solar cell varies from about 12% to 18% with a theoretical maximum of 22%. Several techniques can be used to increase the efficiency, such as by using an optically reflecting back contact metallisation to increase the effective cell thickness and absorption. A back surface field junction may also be introduced, to increase the output voltage or current, by means of a $p^+$ diffusion between the p layer and the back metallisation. A further technique used to improve efficiency is to vary the impurity concentration in a controlled manner across the junction as this increases the carrier collection efficiency.

Silicon cells have a very long operating life provided they are carefully encapsulated. Cells can degrade in performance between 2% and

10% in bright sunlight conditions and under high temperature. They are therefore not suitable for use with concentrating systems which focus the light onto a small area.

There are several variations of the basic silicon solar cell. The heterojunction structures uses a thin p layer of a semiconductor such as tin oxide, indium oxide, cadmium sulphide or gallium phosphide, which is deposited onto an n type silicon base. This system gives a diode between two dissimilar materials and has an efficiency of between 10% and 15%. The top layer has a thickness of only about 1 $\mu$m so that only small quantities of the expensive material are used, the bulk of the cell being made from silicon which is relatively cheap. A non-reflecting surface layer need not now be used so that the heterojunction cell has a good response to short wavelengths and is easy to make.

The Schottky barrier solar cell uses a thin transparent metallic layer of aluminium, chromium or indium–tin oxide on a silicon base. This cell has an efficiency of 8% to 10% and good short wavelength response. It also does not need processing at high temperatures to form the junction.

Although silicon is cheap and readily available the construction method used to form single crystal solar cells is expensive. Research is continuing into methods of reducing this cost, such as by forming the silicon into ribbons or sheets by first casting the silicon in sheets and then crystallising by controlled melting and resolidification. Ribbon forming allows continuous production and automates the process, and by using square ribbon cells rather than the conventional round cells higher packing efficiencies can be obtained when several solar cells are mounted together on a panel.

Amorphous (non-crystalline) silicon has also been used in place of crystalline silicon. Amorphous silicon has a band gap close to the theoretical value of 1.6 eV in sunlight, and a higher absorption coefficient than crystalline silicon. Usually amorphous silicon cells are of the Schottky barrier type.

Thin film silicon cells have been made using evaporation or chemical vapour deposition techniques, which are easy to use in high volume production. The conversion efficiency of these cells is lower than conventional single crystal silicon cells for two main reasons: (i) at least 20 $\mu$m thickness of silicon is used to absorb an appreciable amount of the incident light, and a layer of this thickness cannot easily be produced by evaporation or chemical vapour deposition techniques; (ii) thin films are polycrystalline in structure and the grain boundaries in this structure shorten the life time of the free charge carriers available in the crystal, and so reduce the photocurrent.

### 8.7.3 *Gallium arsenide solar cells*

Gallium arsenide absorbs light near its surface, whereas in silicon this absorption occurs much further down in the material. Therefore a 5 $\mu$m thick gallium arsenide layer will absorb all the visible light whereas a 100 $\mu$m layer of silicon would normally be needed. In silicon the photo-generated carriers need to travel longer distances without recombination so the material needs to be purer than gallium arsenide to avoid defects at which recombination can occur.

The band gap of gallium arsenide is 1.35 eV which is close to the optimum for maximum conversion efficiency in terrestrial sunlight. The cell is also very good for use in concentrating systems. This would normally raise the cell temperature and reduce its photovoltaic efficiency, but the photothermal efficiency would increase, giving a cell with a high overall efficiency up to about 25% for 250 °C to 300 °C and a concentration factor of a 1000 Suns.

The disadvantages of gallium arsenide are that the material is expensive, its supply is limited, and the method currently used to make it from the liquid phase epitaxy is expensive.

### 8.7.4 *Cadmium sulphide solar cells*

The cell consists of a base layer of n type cadmium sulphide (CdS) and a thin surface layer of p type copper sulphide ($Cu_2S$). Both materials are strongly absorbent to light and so a much thinner cell than that using silicon can be made.

The chief merit of the cadmium sulphide cell is that it can be manufactured relatively cheaply. Several techniques are used. In one method cadmium sulphide is vacuum deposited onto a metallised plastic or glass substrate. This is

| Material | Structure | Open circuit voltage (V) | Short circuit current density mA/cm$^2$ | Typical efficiency (%) |
|---|---|---|---|---|
| Silicon | Non-reflecting | 0.6 | 40 | 18 |
| | Heterojunction | 0.5 | 25 | 10 |
| | Schottky barrier | 0.6 | 25 | 10 |
| Cadmium sulphide | Conventional | 0.5 | 25 | 6 |
| Gallium arsenide | Heterojunction | 0.9 | 40 | 25 |
| | Schottky barrier | 0.9 | 20 | 15 |

Fig. 8.20. Comparison of a few solar cells.

then dipped into a hot concentrated solution of copper chloride in water for a few seconds. Copper ions displace cadmium ions at the surface to form a barrier layer of copper sulphide in the cadmium sulphide layer. The copper sulphide layer is very thin which has the advantage that the hole-electron pair generation occurs close to the junction and so increases charge collection. However, since the copper sulphide layer is thin it presents a high resistance to current flow so that short current collection paths are needed. This is usually achieved by a fine gold plated grid structure which is formed onto the copper sulphide layer to act as a contact.

Although the cadmium sulphide cell has a theoretical maximum efficiency of 18% to 25% in practice the cell efficiency is close to 5%. The cells are also unstable and high temperature and high illumination can cause a degradation of the material and a fall in its power output. The cause of degradation is the slow conversion of $Cu_2S$ to $CuS$ and $Cu$. Even under normal conditions the cell loses about 10% of its power output after one year of normal exposure due to the effects of water vapour and oxygen.

Fig. 8.20 compares the basic parameters of some of the more commonly used solar cells.

## 8.8 Special cells

The batteries described in this section are not used in general purpose electronic equipment, but have special applications. Three types of batteries are described, the standard battery, the reserve battery and the radioisotope battery.

The voltage of a battery is determined by the chemical nature of the electrodes and the electrolyte, so that by careful construction it is possible to get a battery with a known terminal voltage. There are many types of standard cells, the best known being the Weston cadmium cell, as shown in fig. 8.21. It consists of a glass tube in the form of an H containing a pool of mercury at the bottom of each leg which makes contact with the metal electrodes inserted into the tube.

At the anode the mercury is amalgamated with 10% cadmium and at the cathode it is in contact with a paste of mercurous sulphate. The electrolyte is a saturated solution of cadmium sulphate. All the materials used in the cell are carefully prepared and controlled, and the output voltage is 1.01864 V at 20 °C with a well defined variation with temperature. The standard cell is used as a voltage standard in electronic equipment. Smaller constructions are used for this application, one version being in the form of a small cylinder 1 cm in diameter and 4 cm long.

Reserve batteries are usually kept in an inactive state for many years. They are activated prior to use in a very short time, usually in less than one second, and are often used in a one-shot application, such as for missiles.

The thermal battery is a reserve battery in which the metal electrodes are separated by an electrolyte consisting of a mixture of alkaline salts with a low melting temperature. The electrolyte is a solid at a storage temperature of −55 °C to +85 °C. Prior to use the electrolyte is melted due to the heat produced in the battery. The electrodes are sheet nickel with one face coated with metallic calcium.

The heating element consists of an oxidiser such as metallic chromate and a metal (zirconium powder) reacting with oxygen. Ignition can occur mechanically or electrically.

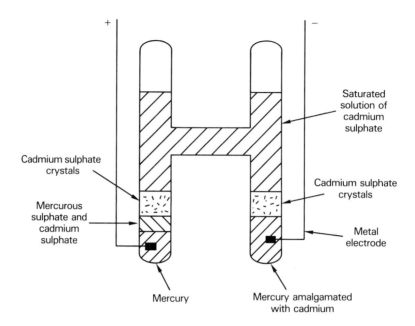

Fig. 8.21. Weston cadmium standard cell.

On ignition a large amount of heat is produced very quickly without any gas. A depolariser is generally added to maintain the reaction for a suitable time.

The operating time is limited by the active materials, such as the electrodes and depolariser, being used up, by the build-up of waste products in the cell, and by the gradual decrease of the initial heat produced in the battery. These effects result in a gradual increase in the cell's internal resistance. The operating time can be designed for a few tens of seconds to a few minutes.

The thermal cell has a low internal resistance and can deliver a large instantaneous power up to about 50 W/cm$^3$. The cell voltage is about 3 V per cell.

Sea batteries are another form of reserve battery which are stored in the dry state and are activated immediately on contact with sea water. They are used at sea in rescue buoys and location markers. There are many types of sea batteries. The silver chloride magnesium cell has an open circuit voltage of 1.5 V and is activated over the temperature range −5 °C to +30 °C. At lower temperatures the initial power output is small but this builds up due to internal heating.

The radioisotope battery is mainly used as a high voltage low current source. It works on the principle that beta rays consist of a stream of high energy electrons. The battery is made by coating an isotope such as strontium 90 between two gold foil strips, and this acts as the emitter. Electrons are collected on an outer aluminium container. High voltages can build up on the cell if time is allowed for charge to accumulate, for example 5 kV is reached after one month. The output current capability at this voltage is about 100 pA.

The radioisotope battery is used in timing and biasing circuits and as power sources for radiation measurement instruments.

# Bibliography

(1) N.C. Wright, *Elementary Semiconductor Physics*, Van Nostrand Reinhold, 1979, 77pp.
(2) A. Bar-Lev, *Semiconductors and Electronic Devices*, Prentice-Hall, 1979, 400pp.
(3) S.K. Ghandhi, *Semiconductor Power Devices*, Wiley, 1977, 329pp.
(4) S. Gage *et al.*, *Optoelectronics Applications Manual*, McGraw-Hill, 1977, 287pp.
(5) Thomas H. Jones, *Electronic Components Handbook*, Reston Publications, 1978, 391pp.
(6) Charles A. Harper, *Handbook of Components for Electronics*, McGraw-Hill, 1977, 1097pp.
(7) John D. Lenk, *Handbook of Electronic Components and Circuits*, Prentice-Hall, 1974, 216pp.
(8) G.W.A. Dummer, *Fixed Resistors*, Pitman, 1967, 242pp.
(9) W. Williams, *Luminescence and the Light Emitting Diode*, Pergamon, 1978, 241pp.
(10) S.A. Stigart, *The J & P Transformer Book* Newnes–Butterworths, 1973, 770pp.
(11) C.J. Richards, *Electronic Display and Data Systems*, McGraw-Hill, 1973.

# Glossary of acronyms

| | | | |
|---|---|---|---|
| APD | Avalanche photodiode | LPE | Liquid phase epitaxy |
| ASCII | American standard code for information interchange | LSA | Limited space charge accumulation |
| BARITT | Barrier injection transit time (diode) | MOSFET | Metal oxide semiconductor field effect transistor |
| BCD | Binary coded decimal | NEP | Noise equivalent power |
| CRT | Cathode ray tube | NTC | Negative temperature coefficient |
| CUJT | Complementary unijunction transistor | PTC | Positive temperature coefficient |
| DIL | Dual-in-line | PTFE | Polytetrafluoroethylene |
| DMOS | Double diffused metal oxide semiconductor | PUT | Programmable unijunction transistor |
| | | PVC | Polyvinyl chloride |
| EHT | Extra high tension | RFI | Radio frequency interference |
| ESR | Equivalent series resistance | SAW | Surface acoustic wave |
| FET | Field effect transistor | SBS | Silicon bilateral switch |
| GTO | Gate turn off | SCR | Silicon controlled rectifier |
| HRC | High rupturing capacity (fuse) | SOA | Safe operating area |
| IDC | Insulation displacement connector | SUS | Silicon unilateral switch |
| IGFET | Insulated gate field effect transistor | TCR | Temperature coefficient of resistance |
| IMPATT | Impact avalanche transit time (diode) | TRAPATT | Trapped plasma avalanche transit time (diode) |
| JFET | Junction field effect transistor | | |
| LASCR | Light activated silicon controlled rectifier | UHF | Ultra high frequency |
| | | UJT | Unijunction transistor |
| LASER | Light amplification by simulated emission of radiation | VDR | Voltage dependent resistor |
| | | VHF | Very high frequency |
| LDR | Light dependent resistor | VMOS | V-groove metal oxide semiconductor |
| LED | Light emitting diode | VPE | Vapour phase epitaxy |
| LEF | Light emitting film | ZIF | Zero insertion force |
| LOC | Large optical cavity | | |

# Glossary of terms

*Active components*: Components made from semiconductor material

*Actuation time*: The time needed for relay contacts to reach their final state after application of coil current.

*Address*: The designation or location to which an electrical signal is applied.

*Alphanumeric display*: A component capable of displaying alphabetic and numeric characters.

*Amplifying gate*: A type of thyristor in which the *gate* current is amplified internally in the device. Also called a *regenerative gate*.

*Antiparallel*: Two components connected in parallel but operating in opposite directions.

*Autotransformer*: A transformer with a single winding for the primary and secondary.

*Bias:* A voltage applied to a component.

*Bipolar*: A semiconductor component in which conduction occurs due to holes and electrons in the same device. See also *unipolar*.

*Bistable*: A component or system which is stable in either one of two modes. See also *monostable*.

*Bounce*: See *contact bounce*.

*Bounce time*: The period over which mechanical contacts of switches and relays will *bounce* before finally remaining closed.

*Carriers*: The holes or electrons which cause conduction in semiconductor devices.

*Characteristics*: Parameters which define how a component will perform under various conditions. See also *ratings*.

*Charge retention*: The ability of a battery or capacitor to hold its charge under specified conditions.

*Chatter*: Variation in *contact resistance* between relay or switch contacts when they first close. It is measured after the *bounce time* and before the *contact resistance* settles to a steady state value.

*Coercive force*: The magnetic field required to overcome the *retentivity* in a magnetic material.

*Coercivity*: See *coercive force*.

*Coherent*: A beam of light in which the individual waves have the same phase relationship.

*Contact*: Metal parts of a switch, relay or connector which make and break the electrical circuit.

*Contact bounce*: The bounce of the metal contacts when they first close causing the electrical circuit to be made and broken at high frequency.

*Contact resistance*: The electrical resistance between the metal contacts of a switch, relay or connector.

*Corona*: Partial discharge in a gas. This can occur in the air spaces contained in transformer insulation.

*Dark current*: The current conducted through a photosensitive device, such as a phototransistor or reverse biased photodiode, when there is no light falling on it.

*Data sheet*: Manufacturer's published information which specifies the performance of an electronic component.

*Depletion mode*: The operation of a field effect transistor such that it is conducting with zero gate voltage. See also *enhancement mode*.

*Depth of discharge*: A measure of the state of charge of a battery. It is defined as the percentage of the rated capacity by which the battery has been discharged.

*Derating*: Reducing one or more of the *ratings* of a component due to an external factor. An example is the reduction in permissible power dissipation of a semiconductor at higher temperatures.

$di/dt$: Rate of change of current.

*Dice*: Plural of *die*.

*Die*: An unencapsulated component, for example a semiconductor before it is put into its package.

*Diffusion*; A technique for introducing *impurities* into the semiconductor. Under the influence of high temperatures these impurity atoms gradually work their way into the crystal lattice of the semiconductor. See also *ion implantation*.

*Dopant*: An additive to a semiconductor designed to give it certain electrical properties. This is also called an *impurity*.

*Dot matrix display*: A display which is made up of an array of individual light sources arranged in a matrix format. The most popular arrangement of the matrix is seven rows and five columns.

$dv/dt$: Rate of change of voltage.

*Dynamic resistance*: The slope of the current–voltage curve of a component at any point, giving the resistance of the device at that point. The curve may be non-linear so that the dynamic resistance of the component will vary along the curve.

*Electroluminescence*: A phenomenon based on the emission of light from a semiconductor under the influence of an electrical field.

*Enhancement mode*: The mode of operation of a field effect transistor in which negligible current flows between source and drain at zero gate *bias*. See also *depletion mode*.

*Epitaxial layer*: A relatively thin layer of semiconductor with a closely defined chemical structure, which is formed on the surface of the semiconductor substrate.

*Equivalent circuit*: An electrical circuit used to simulate the operation of a component.

*Extinction threshold*: The minimum voltage level below which a conducting gas discharge lamp will go off.

*Fall time*: The time needed for a pulse to fall from 90% to 10% of its peak value. See also *rise time*.

*Gain*: The ratio of output current controlled to input controlling current, e.g. collector current to base current of a transistor.

*Gate*: The terminal used to turn on a thyristor or triac, or to turn a gate turn off switch or FET on and off.

*Half life*: Time to reach half the original value of a parameter. For example the half life of a LED is the time for the luminance level to fall to half its original value.

*Heat pipe*: A component for conducting heat from a source to another region where it can be dissipated more easily.

*Hermetic sealing*: Sealing a package to prevent atmospheric contaminants from entering.

*Hybrid*: A component made from two or more different technologies. For example a hybrid relay is made from solid state and electromagnetic parts.

*Impurity*: See *dopant*.

*Insert*: Rigid material contained within a connector housing. The contacts are mounted in this material and are free to move slightly. This helps with alignment and reduces mating forces.

*Integrated circuit*: A circuit which is made as an assembly of electronic elements in a single structure.

*Ion implantation*: A technique for introducing *impurities* into the semiconductor surface by accelerating them to a high velocity and then bombarding them onto the semiconductor. See also *diffusion*.

*Isolation*: The technique used to electrically separate different parts of a system on a semiconductor *die*.

*Leading edge*: The side of a pulse waveform which occurs first in time. The rising amplitude part of a pulse signal. See also *trailing edge*.

*Leakage capacitance*: See *parasitic capacitance*.

*Light dependent resistor*: A resistor whose resistance value varies with the intensity of light falling on it.

*Light emitting diode*: A semiconductor diode which emits light when conducting current.

*Limit switch*: A switch designed to sense the position of an object, and to operate a contact when a set position is reached. There is mechanical contact between the switch and the object being sensed. See also *proximity switch*.

*Liquid phase epitaxy*: A method of forming an *epitaxial layer* on a semiconductor substrate. In one system the constituents of the required compound are dissolved in a suitable material. The substrate is dipped in the melt and cooled to form the epitaxy. See also *vapour phase epitaxy*.

*Long term stability*: A measure of the change in a parameter over a period of time. For example, the long term stability of a capacitor is the change in capacitance value over a long period of time.

*Magneto resistor*: A resistor whose resistance value varies with the strength of the magnetic field in which it is placed.

*Metallisation*: Metal layer formed on top of a semiconductor *die* to provide electrical connection between selected parts of the *die*.

*Monochromatic*: A beam of light which has a single frequency or, in practice, a very narrow band of frequencies.

*Monostable*: A component or system which has only one stable mode, and will return to this mode at the first opportunity. See also *bistable*.

*Multiplex*: To use one channel for several signals.

n *channel*: A field effect transistor in which electrons are the charge carriers between source and drain. See also p *channel*.

n *layer*: A layer of semiconductor which has been doped with n type impurity.

*N-key lockout*: A feature of keyboards which prevents an output signal when two or more keys are depressed simultaneously. See also *N-key rollover*.

*N-key rollover*: A feature of keyboards which enables an operator to depress keys in rapid sequence. See also *N-key lockout*.

*Negative resistance*: A component having a current-voltage curve in which the current decreases as the voltage is increased.

*Negative temperature coefficient*: A parameter which decreases as the temperature is increased.

*Noise equivalent power*: A measure of the noise generated in a device such as a photodiode.

It is the amount of light needed to produce a signal equivalent to the noise level.

*Non-repetitive rating*: The *rating* of the device which should not occur frequently during the life of the device as it may then be destroyed.

*Open circuit voltage*: The voltage at the terminals of a power source, such as a battery, when it is not supplying any load current.

*Operate time*: The time delay between energising the relay coil and the first closure of the contacts.

*Optical coupler*: A component which electrically isolates the input and output signals using light coupling between the two.

*Oxide*: Term often used for the silicon dioxide layer on the semiconductor *die*.

p *channel*: A field effect transistor in which holes are the charge carriers between source and drain. See also n *channel*.

p *layer*: A layer of semiconductor which has been doped with p type impurity. See also n *layer*.

*Parasitic capacitance*: Unwanted capacitance which is unavoidably formed when the system is built. This is also known as *stray capacitance* or *leakage capacitance*.

*Passivation*: A covering on top of a semiconductor *die* which protects it from contaminants. When this protective layer is made from glassy material this process is also known as glassivation.

*Passive component*: A device which is not made from semiconductor material.

*Photometric units*: Units concerned with measurement of visible light, and with the response of the eye to this radiation. See also *radiometric units*.

*Photopic*: Response curve of the human eye at normal illumination levels. See also *scotopic*.

*Planar structor*: A semiconductor *process* in which most of the pn junctions occur at the surface of the *die*.

*Polar diagram*: A plot for a light source showing how the relative intensity of light emission varies with angular displacement.

*Polarised component*: A device which must be used in one way only. For example a polarised electrolytic capacitor must always be operated with one end positive to the other.

*Polarised light*: Light in which wave motion has been modified. In linearly polarised light the wave motion is in one direction only, in a plane perpendicular to the direction of motion.

*Positive temperature coefficient*: A parameter which increases as the temperature is increased.

*Primary cell*: A cell which cannot be recharged and re-used because of irreversible chemical reactions which occur in it when it is producing a current. See also *secondary cell*.

*Process*: The manufacturing steps used in the production of a component.

*Proximity switch*: A switch which operates when it is within a set distance of an object, but where there is no mechanical contact between the switch and the object being sensed. See also *limit switch*.

*Quantum efficiency*: Number of electrons released in a photosensitive component for each photon of incident radiation of a given wavelength.

*Radiometric units*: Units used to measure the radiation of all wavelengths within the optical spectrum. It is therefore concerned with total radiation detection. See also *photometric units*.

*Ratings*: Parameters of a component which must not be exceeded or the device may be destroyed. See also *characteristics*.

*Recombination*: Neutralisation of free holes and electrons in a semiconductor.

*Recovery time*: The time needed for a conducting semiconductor component to recover to its blocking state.

*Recovery voltage*: The voltage to which a battery will recover if it is allowed to rest after a discharge period.

*Rectifier*: A component which will conduct current in one direction only.

*Regenerative gate*: See *amplifying gate*.

*Regulation*: The change in terminal voltage when load is connected to a supply, such as a transformer.

*Reluctance*: The resistance offered by a magnetic material to the flow of magnetic flux.

*Remanence*: The magnetism left in a magnetic material after it has been placed in a magnetic field and the field then removed. The maximum remanence, obtained when the material has been fully magnetised, is called *retentivity*.

*Reserve battery*: A battery which is stored in an inactive state and then activated within a very short time prior to use.

*Retentivity*: See *remanence*.

*Rise time*: The time needed for a pulse to rise from 10% to 90% of its maximum value. See also *fall time*.

*Safe operating area*: The area within the voltage–current characteristic of a power device which defines the region in which the power dissipation of the components is below the maximum value.

*Scotopic*: Response curve of the human eye at low levels of illumination. See also *photopic*.

*Secondary cell*: A cell which can be recharged, and re-used, when it becomes discharged. See also *primary cell*.

*Shelf life*: The time for which a component can be stored without any of its parameters degrading significantly.

*Shielding*: Preventing interference in a component or circuit due to an external field.

*Shorted emitter*: A form of thyristor construction

which enables it to block large rates of change of voltage.

*Skin effect*: The effect in which current flowing at high frequency in a conductor is confined to a thin surface layer. This causes an increase in resistance at high frequencies.

*Snap action switch*: A switch whose contacts are operated by a spring so that it rapidly makes and breaks a circuit. The operate time is independent of the speed with which the actuating arm of the switch is moved.

*Steradian*: A measure of solid angle. It is defined as the solid angle formed at the centre of a sphere by an area, equal to the square of the radius, on the surface of the sphere.

*Stray capacitance*: See *parasitic capacitance*.

*Surge rating*: The *rating* of a component when it is carrying current for a very short time. Usually this is higher than its continuous *rating*.

*Temperature coefficient*: The rate of change of a parameter with temperature. For example the temperature coefficient of resistance is the change of resistance value with temperature.

*Thermal resistance*: A measure of the resistance of a component to the flow of heat. One unit of thermal resistance is that which causes a temperature difference of 1 °C when 1 W of heat flows through it.

*Thermoplastic*: A material which can be melted and solidified on cooling. It can be remelted and resolidified many times without any appreciable change in properties. See also *thermoset*.

*Thermoset*: A material which undergoes a chemical change while being melted. Once solidified it cannot be remelted. See also *thermoplastic*.

*Trailing edge*: The side of a pulse which occurs after the leading edge. The falling amplitude part of a pulse signal. See also *leading edge*.

*Transient voltage*: A surge of voltage of short duration and irregular waveform.

*Unipolar*: A semiconductor component in which conduction occurs either due to holes or electrons but not both at the same time. See also *bipolar*.

*Vapour phase epitaxy*: A method for forming an *epitaxial layer* on a semiconductor substrate. In one system the epitaxy material is carried in a stream of gas and deposited on the surface of the substrate. See also *liquid phase epitaxy*.

*Viewing angle*: A measure of how well a display can be seen when viewed from a side. It is defined as the angle at which the luminous intensity of the display is half the value observed when it is viewed head on.

*Voltage coefficient*: The rate of change of a parameter with voltage. For example the voltage coefficient of resistance is the change of resistance value with temperature.

*Washout*: An effect which makes it very difficult to read some displays in strong direct light, usually due to unwanted reflections within the display.

# Index

Absolute linearity, 59
Absorption hologram, 51
Acoustic noise, 87
Actuation force, 108, 111
Actuation magnet, 113
Actuation time, 91
Address, 46
Ageing
  of capacitor, 79
  of crystal, 136
  of magneto resistor, 106
Alkaline manganese cell, 149
Alphanumeric display, 40
Amorphous material, 140, 143, 162
Amplifier, 137
Amplifying gate, 22
Anisotropic phase, 43
Anode, 3, 69, 71
Antiparallel, 22
APD, 36, 50
Aperture plate, 46
Arcing
  in fuse 118
  in relay, 92
  in switch, 108
Armature, 89, 93
Armature travel, 92
ASCII, 115
AT cut, 134
Attenuation, 49
Autotransformer, 87
Avalanche breakdown, 9, 11
Avalanche conduction, 3
Avalanche photodiode, 36, 50

Back projection display, 48
Backlash, 59
Backswing, 86
Band gap, 30, 160
Bandwidth, 86, 135
BARITT diode, 9
Battery, 145
Battery charging, 156
BCD, 110, 115
Bellows socket, 128
BH loop, 86

Bias
  magnet, 95, 113
  voltage, 6, 38
Bidirectional switch, 112
Bimorph, 138
Binary coded decimal, 110, 115
Bipolar capacitor, 71
Bipolar phototransistor, 37
Bipolar transistor, 11
Birefringent, 44
Blow moulding, 127
Bounce, 95
Bounce time, 90
BT cut, 134
Bubble keyswitch, 114
Bulk glass, 140
Bulk resistor, 106
Bulk semiconductor, 1
Burden, 86,
Button cell, 149, 151, 154

$C$ rate, 146
Cadmium sulphide solar cell, 162
Calchogenide, 140
Candela, 29
Capacitance
  diffusion, 6
  load, 135
  parasitic, 37, 39
  stray, 18, 57
Capacitive key, 115
Capacitor, 66
Carbon film, 114
Carbon resistor, 55, 61
Carbon potentiometer, 61
Carbon–zinc cell, 147
Carrier, 12
  lifetime of, 6, 24, 38
  mobility of, 6, 101
Catalyst, 159
Catalytic reformer, 159
Cathode, 3, 71
Cathode ray tube, 47
Cathode sputtering, 45
Ceramic capacitor, 76
Cermet potentiometer, 61

Change-over switch, 109
Characteristics
  of capacitor, 70, 78
  of connector, 124
  of crystal oscillator, 135
  of diode, 4
  of fuse, 117
  of Hall effect device, 102
  of LED, 32
  of magneto resistor, 106
  of potentiometer, 58
  of relay, 89
  of resistor, 53
  of selenium rectifier, 144
  of switches, 107
  of thyristor, 19
  of transformer, 85
  of transistor, 11
  of varactor diode, 6
Charge acceptance, 146
Charge carrier, 12, 38
Charge reserve, 154
Charge retention, 146
Charge storage, 143
Charge voltage, 146
Chemical deposition, 35, 56
Cholesteric, 43
Clapper relay, 93
Coded switch, 110
Coercive force, 85
Coercivity, 84
Coherent, 28, 33, 51
Collimated light, 49
Complementary UJT, 27
Concentration factor, 162
Conduction band, 30
Conductive film keyswitch, 114
Conductive plastic potentiometer, 61
Conformal coating, 78
Connector, 124
Constant current charging, 146, 155
Constant voltage charging, 157
Contact arcing, 92, 108
Contact bounce, 90, 96, 108
Contact chatter, 90, 97

Contact deformation, 124
Contact derating, 92
Contact force, 92, 108, 124
Contact life, 97
Contact material, 92, 108, 125
Contact noise, 80, 97
Contact protection, 92
Contact resistance
  of capacitor terminals, 68
  of connector, 124
  of potentiometer, 60
  of relay, 90, 92, 97
  of switch, 108, 115
Contact teasing, 109
Contact welding, 92, 109
Contrast, 140
Contrast ratio, 43
Corona, 85, 125
Coupling coefficient, 138
Critical angle, 33, 50, 139
Cross point switch, 115
Cross talk, 38
CRT, 47
Crystal, 1, 134
Crystal can relay, 94
Crystal drag, 136
CUJT, 27
Curie brothers, 132
Curie point, 84, 97
Curie temperature, 63, 97, 132
Curing temperature, 79
Current crowding, 124
Current derating, 119, 125
Current let through, 117
Current limit, 157
Current overload, 116
Current ratio, 87
Current tap, 58
Current transfer ratio, 39
Current transformer, 86
Curve factor, 160
Cut-off region, 24
Cut-off voltage, 146
CV product, 69, 73
Cylindrical cell, 149, 154
Czochralski growing, 1, 31

Dark current, 36, 38
Darlington transistor, 16
De-ageing, 79
Decay time, 48
Delay line, 137
Demagnetisation curve, 83
Demagnetising field strength, 84
Depletion mode, 12
Depletion region, 12
Depolariser, 164

Depoling, 133, 139
Depth of discharge, 146
Detectivity, 36
Diac, 27
Diaphragm relay, 95
Die, 5, 37, 77, 121
Dielectric, 66, 69
  high $K$, 78
Dielectric absorption, 68
Dielectric anisotropy, 44
Dielectric breakdown, 53, 67
Dielectric constant, 43, 79, 81, 85, 133
Dielectric delamination, 77
Dielectric displacement, 133
Dielectric loss, 66, 70
Dielectric resin, 74
Dielectric strength, 67, 85, 90, 121
Differential force, 108
Diffusion, 2, 38
Diffusion capacitance, 6
Diffusion depth, 36
DIL, 110
Diode capacitance, 5
Diode noise, 36
Direct energy gap, 31
Display, 40
  piezoelectric, 139
Dissipation factor
  of capacitor, 66, 70, 74, 76
  of thermistor, 63
Divergence angle, 49
DMOS, 18
Dopant, 5
Doped silicon, 1
Doping
  gold, 24
Doping level, 8
Doping profile, 6
Dot matrix display, 41, 46
Double heterojunction laser, 34
Drift, 60, 63
Drift current, 36
Dry cell, 147
Dry circuits, 92, 108, 114
Dual-in-line socket, 131
Dual-in-line switch, 109
d$v$/d$t$ effect, 38

Eccentricity, 59
ECMA, 115
Eddy current, 88, 113
Efficiency
  of heat sink, 119
  of laser, 49
  of LED, 30, 43
  of photodiode, 36

  of relay, 95, 96
  of solar cell, 162
  of transformer, 85
EHT, 143
Elastic compliance, 133
Elastic deformation, 124
Electrical axis, 134
Electrical conductivity, 125
Electrical displacement. 133
Electrical isolation, 121
Electrical overtravel, 58
Electrical polarisation, 132
Electrical travel, 58
Electro-optics, 139
Electrochromic cell, 48
Electroform, 143
Electroluminescent, 30
Electroluminescent diode, 32
Electroluminescent display, 48
Electrolyte
  in capacitor, 69
  in cell, 146, 159
Electrolytic capacitor, 69
Electromagnetic, 89
Electromagnetic radiation, 28
Electromagnetic shield, 86
Electromagnetic spectrum, 28
Electromechanical coupling coefficient, 133
Electron bombardment, 45
Electron carrier, 12
Electron charge, 101
Electron conduction, 72
Electron drift velocity, 9
Electron emission, 23
Electron injection, 48
Electron mobility, 9, 104
Electrophoretic display, 48
Electrostatic shield, 86, 95
Emitter diode, 24
Emitter lip resistance, 22
End play, 59
End resistance, 58
Energy, 28
Energy band, 30
Energy density of cell, 146
Energy gap, 30, 101
Energy let through, 119
Energy product, 84
Energy stored in capacitor, 68
Energy stored in magnet, 83
Enhancement mode, 12
Epi-base, 14
Equivalent circuit
  for capacitor, 66
  for heat sink assembly, 120
  for quartz crystal, 135

for reed relay, 94
for resistor, 52
Equivalent series resistance, 66
ESR, 66, 69
Etched foil capacitor, 69
Exothermic, 154
Extinction threshold, 46
Extrusion, 127

Fail-safe (switch), 111
Fall time, 12
Fast recovery, 5, 24
Ferrite core, 86
Ferrite Hall device, 102
Ferroeletric, 132
Ferroelectric display, 140
Ferromagnetic, 132
FET, 12, 16, 37
Fibre optic cable, 49
Fibre optic guide, 48
Field effect display, 43
Field effect phototransistor, 37
Field effect transistor, 12
Field strength, 84
Figure of merit, 7, 64
Filament display, 48
Filament lamp, 48
Fill factor, 160
Filter
    crystal, 136
    optical, 44
Flashover, 85
Flat cell, 151
Fluorescence, 48
Fluidised bed, 78
Flux density, 84, 89
Flux fringing, 89
Foil and film capacitor, 74, 79
Forbidden gap, 30
Forced cooling, 80, 120
Fourier transform 51
Fraunhofer hologram, 51
Free position (switch), 105, 110
Frequency multiplier, 6
Frequency resistor characteristic, 52
Frequency spectrum, 48
Fresnel hologram, 51
Fuel cell, 158
Fuse, 116

Gain, 9, 11
Gallium arsenide, 1, 6, 18, 102
Gallium arsenide solar cell, 162
Gas discharge, 30
Gas discharge display, 45
Gas discharge ignition, 137
Gas discharge phase, 102

Gate, 12
Gate current, 21
Gate turn-off, 23
Gelled electrolyte, 73
Glass, 49, 55, 140
Glow discharge, 92
Glow lamp, 30
Graded fibre, 50
Green ceramic, 77
GTO, 23
Gunn effect, 8

Half life, 32
Half power point, 41
Hall angle, 104
Hall constant, 102
Hall effect, 100
Hall key, 113
Hall switch, 115
Halogen cycle, 30
Hard materials, 132
Harmonics, 6
HCR fuse, 117
Heat pipe, 18, 122
Heat sink, 119
Heat sink compound, 122
Heterojunction laser, 34
High voltage connector, 130
Holding current, 19
Hole–electron pair, 37
Holography, 51, 140
Homojunction laser, 34
Hot carrier diode, 6
Hot spots, 55, 123, 125
Housing (connector), 128
Hybrid circuit, 81
Hybrid integrated optics, 50
Hybrid LED display, 43
Hybrid relay, 98
    Schottky diode, 6
Hypertac contact, 128
Hysteresis
    in magnetic materials, 132
    in photoconductor, 35
    in switch, 113
Hysteresis loop, 140
Hysteresis loss
    in capacitor, 67
    in inductor, 88

Idle power, 65
IGFET, 12
Ignition threshold, 45, 46
Illumination, 29
IMPATT diode, 9
Impurities, 1
Incandescence, 48

Independent linearity, 59
Inductive proximity switch, 113
Inductive zero component, 103
Inductor, 88
Industrial fuse, 117
Infrared response, 38
Injection laser, 49
Injection moulding, 127
Insert, 125
Insertion force, 127
Insulation breakdown voltage, 67
Insulation displacement connector, 129
Insulation material, 126
Insulation resistance, 66, 76, 79, 95, 125
Integrated circuit, 137
Integrated circuit socket, 131
Integrated optics, 50
Integrated sensor, 38
Interdigitated gate, 22
Interference colours, 45
Interference pattern, 51
Intermittent failure (switch) 108
Internal reflection, 50
Intrinsic region, 7
Intrinsic stand-off ratio, 25
Ion, 45, 47, 62, 146
Ion conduction, 72
Ion implantation, 3
Ionisation voltage, 74
Irradiance, 29, 36
Iso-$Q$ curves, 88
Isolation voltage, 39
Isolation washer, 121
Isotropic phase, 43

JFET, 12
Johnson noise, 54, 57, 60
Jump-off voltage, 58
Junction capacitance, 36
Junction loss, 161

Keyboard, 113
Keyswitch, 113
Keytop, 116

LASCR, 37
Laser, 33, 49, 55, 140
Lasing threshold, 49
Latent heat, 123
LDR, 34
Lead bending stress, 54
Lead–acid cell, 152
Leakage current, 3, 11, 65, 73, 99, 125
Leakage factor, 84

Leakage inductance, 87
Leakage region (of VDR), 63
Leakage resistance, 66, 149
Leclanché cell, 147
LED, 30, 40, 49, 99
LED display, 41
LEF, 48
Life
   of battery, 146, 156
   of capacitor, 71, 73, 80
   of display, 41
   of glass switch, 142
   of LED, 32
   of liquid crystal, 44
   of magnet, 85
   of mechanical switch, 108, 110, 114
   of potentiometer, 60
   of relay, 93
   of resistor, 57
Light, collimated, 49
Light, polarised, 28
Light absorption, 48
Light activated SCR, 37
Light dependent resistor, 34, 61
Light emitting diode, 30, 49
Light emitting film, 48
Light emitting diode display, 41
Light pipe, 43
Light velocity, 28
Limit switch, 110
Linear array, 38
Linearising resistor, 102
Linearity, 59
Liquid crystal display, 43, 139
Liquid electrolyte, 73
Liquid phase epitaxy, 31, 162
Lithium cell, 150
Litz wire, 88
Long term stability
   of capacitor, 68
   of resistor, 53
Lorentz forces, 104
Loss angle, 66
Loss factor, 84
Loss tangent, 88
LPE, 31
LSA, 8
Luminance, 29
Luminous efficiency, 32
Luminous energy, 29
Luminous exitance, 29
Luminous intensity, 29, 41, 43
Luminous power, 29

Magnetic components, 83
Magnetic core, 89

Magnetic materials, 84
Magnetic proximity switch, 113
Magnetic shielding, 95
Magnetisation curve, 83
Magneto resistor, 61, 104
Magnetostriction, 86, 95
Magnets 83, 113
Magnitude error, 87
Maintaining voltage, 47
Make contact, 89
Mating force, 125
Mechanical axis, 134
Mechanical resonance, 133
Mechanical travel, 59
Memory cell, 156
Memory display, 47
Memory glass, 141
Memory, optical, 140
Mercuric oxide cell, 149
Mercury-wetted relay, 96, 114
Mesa, 6
MESFET, 18
Metal film resistor, 56
Metal glaze resistor, 56
Metal oxide resistor, 55
Metallised film capacitor, 74, 79
Mica capacitor, 79
Microscopy, 51
Modulation, 49
Mono-mode (keyboard), 115
Monochromatic, 28, 33
Monolithic, 50
Monolithic capacitor, 77
Monostable, 141
MOS, 44
MOSFET, 12
Moving armature relay, 93
Multi-mode (keyboard), 116
Multilayer capacitor, 77
Multimorph, 138
Multiplexing, 38, 44

n channel, 12
n impurity, 1
$N$-key lockout, 116
$N$-key rollover, 116
Negative resistance diode, 8
Negative resistance effect, 24
Negative temperature coefficient, 53
Nematic, 43
NEP, 36
Nickel—cadmium cell, 153
Nixi tube, 45
Noise
   capacitor, 68, 80
   diode, 6, 8
   photodiode, 36

   potentiometer, 60
   reed relay, 95
   resistor, 53, 56
   transformer, 86
Noise equivalent power, 36
Non-inductive winding, 57
Non-linear resistor, 61
Non-polar capacitor, 71
NTC thermistor, 61
Numeric indicator tube, 45
Numerical aperture, 50

O-ring, 60
Offset voltage, 37
Ohm's law, 52
Open circuit voltage, 145
Operate position (switch), 107, 110
Operate time (relay), 90, 95
Operating force (switch), 107
Optical axis, 134
Optical communication, 49
Optical conversion efficiency, 43
Optical coupler, 38
Optical detector, 34, 38
Optical memory, 140
Optical modulator, 140
Optical oscillator, 33
Optical sources, 29, 38
Optical switch, 115
Optical terminology, 28
Oscillator, 24, 113, 134, 137
Oxide, 3, 55
Oxide crystallisation, 74
Oxide film, 71
Oxide glass, 140

p channel, 12
p impurity, 1
Paper capacitor, 74
Parallel antiresonance, 135
Paramagnet, 97
Parasitic capacitance, 37, 39
Partial failure, 146
Passivated structure, 10
Peak point, 24
Permanent magnet, 83, 94
Permeability, 88, 97
Permittivity, 66, 74, 133, 138
Phase error, 87
Phase hologram, 51
Phosphor, 48
Photo Darlington, 37, 40
Photo-detector, 34
Photo-emissive devices, 38
Photoconductive devices, 34, 140
Photodiode, 35, 40

# Index

Photolithography, 3, 137
Photometric unit, 28
Photomultiplier tube, 38
Photon, 31, 34
Photopic, 28
Photoresistive device, 34
Photothyristor, 37, 40
Phototransistor, 36, 40
Photovoltaic, 35, 160
Pinch-off, 12
Planar structure, 24
    diffused, 5
    epitaxial, 14
Planck's constant, 31
Plasma panel, 46
Plastic capacitor, 74
Plastic deformation, 124
PLZT, 140
Polar diagram, 33
Polarised capacitor, 71
Polarised cell, 147
Polarised connector, 124
Polariser, 43, 139
Positive drive (switch), 109
Positive temperature coefficient, 53
Pot core, 89
Potentiometer, 57
Powder core, 89
Power capacitor, 80
Power derating, 53
Power dissipation, 3, 52, 62, 67, 85
Power factor, 66, 85, 87, 89
Power FET, 16
Power rating, 58
Power relay, 93
Power sources, 145
Power transformer, 86
Pre-aged, 79
Pre-arcing time, 118
Pre-travel, 107
Press pak package, 5, 24
Primary cell, 145, 146
Primary aperture, 46
Printed circuit board, 57, 110
Printed circuit board connector, 127, 130
Programmable UJT, 27
Proximity switch, 112
PTC thermistor, 63
PTFE, 130, 151
Pull-in force (relay), 91
Pullability (of crystal), 136
Pulse transformer, 86
Push button switch, 109
PUT, 27
PVC, 50
Pyroelectric, 132, 140

$Q$ factor, 7, 81, 88, 113, 134
Quantum efficiency, 30, 32, 36
Quartz crystal, 134

Radial play, 59
Radiance, 29
Radiant energy, 29
Radiant exitance, 29
Radiant intensity, 29
Radiant power, 29
Radio frequency interference, 49, 100
Radiometric unit, 28
Ratings
    capacitor, 67
    diode, 4
    resistor, 52
    thyristor, 19
    transistor, 11
Read diode, 9
Reconstituted mica, 79
Recovery time, 5
Rectification, 6
Rectifier, 19
Reed keyswitch, 113
Reed relay, 113
Reflection coefficient, 33
Reflective display, 43
Refractive index, 33, 43, 139, 160
Regenerative gate, 22
Regulation, 85, 87
Relative permittivity, 71
Relay, 89
Reliability
    of capacitor, 73
    of connector, 128
    of relay, 99
    of resistor, 57
Reluctance, 84
Remanence, 84
Reserve battery, 163
Residual loss, 88
Resin dielectric, 74
Resistive zero component, 103
Resistor, 52
Resolution of potentiometer, 59, 61
Resonance (in crystal), 136
Resonant frequency, 69, 128
Responsivity, 36, 38
Reserve leakage current, 5
Reserve recovery, 5, 143
Reserve voltage, 3
RFI, 49
Ribbon wire, 128
Ripple current, 67
Rise time, 12, 48, 74, 87
Rocker switch, 109

Rotary switch, 109, 110
Rotor, 81
Running torque, 59
Rupturing time (fuse), 118

Safe operating area, 15
Saturation region, 11
SAW, 137
SBS, 27
Scan anode, 46
Schottky diode, 6, 36
Schottky solar cell, 162
Scotopic vision, 28
Screen printing, 141
Sea battery, 164
Sealed switch, 110
Seating time, 91
Second breakdown, 15
Secondary cell, 145, 152
Selenium, 143
Self-demagnetisation, 85
Self-depolarisation, 147
Self-discharge, 146, 153
Self-healing capacitor, 72, 74
Self-heating, 62
Self-inductance, 79
Self-scan, 46
Semiconductor production, 1
Semiconductor protection fuse, 117
Sensitivity
    of phototransistor, 37
    of relay, 92, 94, 97
    of thermistor, 63
Separator (in cell), 153
Series resonance (in crystal), 135
Shaft run out, 59
Shelf life, 71, 147, 153
Shorted emitter, 21
Signal to noise ratio, 51
Silica, 3
Silicon bilateral switch, 27
Silicon defects, 38
Silicon display, 41
Silicon solar cell, 161
Silicon unilateral switch, 27
Solid electrolyte, 159
Silver elastomer film, 114
Silver oxide cell, 150
Silver-zinc cell, 156
Sintering, 35
Skin effect, 88
Slide switch, 109
Slope efficiency, 49
Smectic, 43
Smoothing choke, 89
Snap-action switch, 109
Snell's law, 50

## 176  Index

SOA, 15
Socket, 127
Soft materials, 132
Solar cell, 159
Solid angle, 33
Solid electrolyte, 71
Solid state relay, 98
Solid tantalum capacitor, 71
Space charge region, 36
Span ratio, 86
Specific resistance, 72
Spectral response, 35, 38
Speed (of display), 44
Sputtering, 3, 45, 46
Stabilising field, 84
Stability
    of capacitor, 80
    of Hall device, 102
    of inductor, 89
    of oscillator, 137
    of resistor, 55
Stabistor, 11
Standard cell, 163
Starting torque, 59
Static resistance (of VDR), 65
Static wicking height, 123
Stator, 81
Step fibre, 50
Step-recovery diode, 7
Steradian, 29
Storage life, 71
Stray capacitance, 57
Stud package, 5, 24
Superheating, 123
Surface acoustic wave, 137
Surface leakage, 76
Surge current, 4, 55, 92
Surge voltage, 67
SUS, 27
Switch, 107
Switching curves (of switch), 114
Switching glass, 140
Switching temperature, 63, 97
Switching time, 142
Switching waveform, 11
Symbol
    for CUJT, 26
    for diode, 3
    for JFET, 13
    for MOSFET, 14
    for relay, 90
    for SBS, 26
    for SUS, 26
    for thyristor, 19
    for triac, 23
    for UJT, 25
    for zener diode, 10

Tantalum electrolytic capacitor, 71
Tantalum foil capacitor, 71
Tap, 58
Tap ratio, 87
Taper charging, 157
TCR, 53, 56
Telephone relay, 94
Temperature coefficient
    of capacitor, 68, 81
    of expansion, 63, 102
    of frequency, 134
    of Hall voltage, 101, 102
    of inductance, 88
    of resistance, 53
    of zener diode, 11
Temperature controlled oscillator, 136
Temperature controlled oven, 135
Temperature cycling, 84
Temperature factor, 88
Temperature sensor, 98, 142
Thermal battery, 163
Thermal capacitance, 120
Thermal circuit, 119
Thermal conductivity, 9, 102, 121
Thermal efficiency, 158
Thermal energy, 160
Thermal grease, 122
Thermal impedance, 120
Thermal noise, 54
Thermal relay, 97
Thermal resistance, 5, 120
Thermal voltage, 95
Thermomagnetic switch, 97
Thermistor, 61
Thermoplastic, 55, 126
Thermoset, 126
Thick film, 46
Thick hologram, 51
Thin film battery, 159
Thin film resistance, 46
Thin film solar cell, 162
Threshold current, 86
Threshold lasing, 49
Threshold light level, 38
Threshold switch, 141
Threshold voltage, 44, 141
Thumbwheel switch, 110
Thyristor, 19, 99
Time constant, 68, 76
Time delay, 91, 97, 141
Time delay relay, 100
Time lag, 68
TO-5, 24
TO-5 relay, 94
Toggle switch, 109
Tolerance, 53, 60

Toroid, 86, 89
Torque, 59, 80
Total travel, 59
Total travelled position, 107, 110
Transfer curve, 12
Transfer moulding, 127
Transfer ratio, 86
Transformer, 85
Transient energy, 65
Transient thermal analysis, 120
Transient time, 8
Transient voltage suppressor, 143
Transistor, 11, 99
Transmissive display, 43
TRAPATT diode, 9
Triac, 22, 99
Trickle charge, 157
Trigger device, 24
Trigger diode, 27
Trimmer capacitor, 80
Trimmer potentiometer, 57
Triple diffused, 14
Tunnel diode, 8
Turn-off gain, 24
Turn-off time, 21, 38, 44
Turn-on time, 21, 44
Turns ratio, 87
Twisted nematic display, 43

UJT, 24
Unidirectional switch, 112
Unijunction transistor, 24
Unipolar transistor, 12
Unmating force, 125
Upturn region (of VDR), 64

Vacuum deposition, 3, 35, 140, 162
Vacuum evaporation, 56
Vacuum forming, 127
Valley point, 24
Vapour phase epitaxy, 31
Varactor diode, 6
Variable capacitor, 80
Variable conductance, 124
Variable resistor, 57
Varistor, 64
VDR, 61, 63
Vertically polarised, 43
Viewing angle, 41
Virtual image, 51
Viscosity, 44
VMOS, 16
Voltage, thermal, 95
Voltage coefficient, 53
Voltage controlled oscillator, 136
Voltage dependent resistor, 61, 63
Voltage derating factor, 67

Index 177

Voltage divider, 60
Voltage overshoot, 86
Voltage ratio, 59
Voltage reference diode, 10
Volume hologram, 51
Volume resistivity, 67
VPE, 31

Wall charge, 47
Washout, 43, 139
Wave soldering, 129
Waveguide, 50
Weston cadmium cell, 163
Wick (heat pipe), 123
Wiper, 60
Wirewound potentiometer, 61
Wirewound resistor, 56
Withdrawal force, 128
Working fluid, 122
Wound capacitor, 74

Zener diode, 10, 27
Zero width tap, 59
Zero insertion force, 125
Zero insertion force connector, 131
ZIF, 131
Zinc chloride cell, 148
Zinc–air cell, 151
Zone levelling, 1